# Global Forest Carbon

This book addresses the major policy, economic and financial issues encountered in global forest carbon.

The global forest sector is expected to play a major role in achieving the Paris Agreement's temperature targets. Therefore, there is an urgent need to explore practical and promising solutions to the challenges facing carbon accounting and policy assessment as the global community undertakes forest sector actions—including the widely known REDD+ initiative. This book demonstrates how vital it is that we identify appropriate perspectives and formulate approaches to address these challenges in an integrated and effective manner. In doing so, it addresses many of the major issues, including the differential potentials for carbon sequestration within various forest ecosystems as well as for storage within a variety of harvested wood products, the joint production of timber and carbon, and the measurement and impact of forest carbon offsets and credits, results-based payments, and other nationally determined contributions centered differences as well. The book examines regional and country-level case studies from across the world and draws on the author's decades of experience working on forest policy and with the forest sector. Overall, this book highlights the technical and policy issues regarding forest sector carbon emission and removal to build useful perspectives, frameworks, and methods for addressing these issues successfully in the future. It advances the knowledge frontiers of global forest carbon policy, economics and finance as well as the ability to assess the effectiveness, efficiency, and equity of forest climate solutions.

This book is essential reading for professionals and policymakers working at the intersection of forest policy, carbon storage and climate change, as well as students and researchers in the fields of forestry, natural resource management, climate change and nature-based solutions.

**Runsheng Yin** is a Professor of Forest Ecosystem Economics and Policy at Michigan State University, USA. Over the past three decades he has conducted research projects evaluating ecosystem restoration programs, analysing forest tenure reforms, accounting for and assessment of forest carbon, and exploring rural sustainable development. He has published over ninety peer reviewed papers and served as an editor-in-chief for the international journal of *Forest Policy and Economics*.

"*Global Forest Carbon*...is one of the first books that tackle issues of forest carbon policy, economics, and finance in a coherent, practical, and timely manner. I would recommend it to my students and colleagues as well as other scholars and policymakers who are interested in forest sector climate solutions."

**Shilong Piao**, *Professor, Institute of Carbon Neutrality, Peking University*

"*Global Forest Carbon: Policy, Economics and Finance* is an excellent resource on the role of forest sector in addressing climate change. The author uses analytical, critical, and comprehensive lens and global cases to present an exemplary synthesis of the processes, principles and practices for the reduction and removal of carbon emissions."

**Shashi Kant**, *Professor, Institute for Management & Innovation, University of Toronto*

"Empowering and enlightening! *Global Forest Carbon*...illuminates the complex links between carbon markets, sustainable finance, and forest conservation, providing a comprehensive view of forest carbon management. A persuasive read that equips readers to champion nature-centered climate remedies by fostering profound insight and capability in tackling the crucial challenges of policy implementation."

**Wan-Yu Liu**, *Professor, Department of Forestry, National Chung Hsing University, Taiwan*

"The questions *Global Forest Carbon*...addresses are key to inform sound forest sector policies in implementing the Paris Agreement, providing valuable insights on the complex topics of forest carbon economics and finance. It is a most welcome contribution to the ongoing efforts on climate change adaptation and mitigation."

**Monica Gabay**, *Professor, National University of San Martin, Argentina; Division 9 Coordinator, International Union of Forest Research Organizations*

# The Earthscan Forest Library

**Series Editorial Advisers**: **John L. Innes**, *University of British Columbia, Canada*; **John Parrotta**, *US Forest Service - Research & Development, USA*; **Jeffrey Sayer**, *University of British Columbia, Canada; and* **Carol J. Pierce Colfer**, *Center for International Forestry Research, USA*.

This series brings together a wide collection of volumes addressing diverse aspects of forests and forestry and draws on a range of disciplinary perspectives. It is aimed at undergraduate and postgraduate students, researchers, professionals, policy-makers and concerned members of civil society.

**Masculinities in Forests**
Representations of Diversity
*Carol J. Pierce Colfer*

**Adaptive Collaborative Management in Forest Landscapes**
Villagers, Bureaucrats and Civil Society
*Edited by Carol J. Pierce Colfer, Ravi Prabhu, and Anne M. Larson*

**Sacred Forests of Asia**
Spiritual Ecology and the Politics of Nature Conservation
*Edited by Chris Coggins and Bixia Chen*

**The Cultural Value of Trees**
Folk Value and Biocultural Conservation
*Edited by Jeffrey Wall*

**Responding to Environmental Issues through Adaptive Collaborative Management**
From Forest Communities to Global Actors
*Edited by Carol J. Pierce Colfer and Ravi Prabhu*

**Global Forest Carbon**
Policy, Economics and Finance
*Runsheng Yin*

For more information about this series, please visit: www.routledge.com/The-Earthscan-Forest-Library/book-series/ECTEFL

# Global Forest Carbon
Policy, Economics and Finance

**Runsheng Yin**

Designed cover image: © Getty Images

First published 2024
by Routledge
4 Park Square, Milton Park, Abingdon, Oxon OX14 4RN

and by Routledge
605 Third Avenue, New York, NY 10158

*Routledge is an imprint of the Taylor & Francis Group, an informa business*

© 2024 Runsheng Yin

The right of Runsheng Yin to be identified as author of this work has been asserted in accordance with sections 77 and 78 of the Copyright, Designs and Patents Act 1988.

All rights reserved. No part of this book may be reprinted or reproduced or utilised in any form or by any electronic, mechanical, or other means, now known or hereafter invented, including photocopying and recording, or in any information storage or retrieval system, without permission in writing from the publishers.

*Trademark notice*: Product or corporate names may be trademarks or registered trademarks, and are used only for identification and explanation without intent to infringe.

*British Library Cataloguing-in-Publication Data*
A catalogue record for this book is available from the British Library

ISBN: 978-1-032-56654-2 (hbk)
ISBN: 978-1-032-56536-1 (pbk)
ISBN: 978-1-003-43665-2 (ebk)

DOI: 10.4324/9781003436652

Typeset in Times New Roman
by Taylor & Francis Books

# Contents

*List of illustrations* ix
*Acknowledgements* xi
*Foreword* xiii
*Foreword* xv

Introduction 1

1 A primer on forest carbon policy and economics under the Paris Agreement 9

2 Understanding and undertaking jurisdictional approaches to governing forest climate actions 25

3 Forest carbon accounting and assessment: Alternative perspectives and methodologies 41

4 Outcomes of different frameworks in forest carbon accounting at the local level 61

5 The forest sector's potential role in national decarbonization: China as an example 78

6 Carbon leakage induced by forest products trade: Outcomes and implications of alternative accounting practices for China 97

7 Carbon finance and funding for forest sector climate solutions 113

8 Risk, uncertainty, and social discounting in coping with climate change 131

| | | |
|---|---|---|
| 9 | Evaluating the evaluated impacts of forest carbon programs and projects | 149 |
| 10 | Summary and outlook | 167 |
| | *Index* | 188 |

# Illustrations

**Figures**

| | | |
|---|---|---|
| 5.1 | Alternative trajectories of carbon emissions projected by the GCAM-TU and the IPAC model | 82 |
| 5.2 | Alternative trajectories of projected forest area and stock volume under the conventional case (I) and the accelerated case (II) | 84 |
| 5.3 | Alternative pathways (SSPs) of C storage in HWPs | 85 |
| 6.1 | China's consumption- vs. production-based $CO_2$ emissions | 98 |
| 6.2 | China's trade balance of major forest products (in 2017 constant price) | 106 |
| 6.3 | Estimated coefficients of $CO_2$ emissions for China's forest products industry from 2000 to 2015 | 107 |
| 6.4 | Different measurements of $CO_2$ emissions for China's forest products industry from 2000 to 2014 | 108 |
| 6.5 | $CO_2$ emissions with 21 trading partners of forest products in 2014 | 108 |
| 6.6 | $CO_2$ emissions embodied in China's forest products trade in 2000 and 2014 (see country definitions beneath Figure 6.5) | 109 |

**Tables**

| | | |
|---|---|---|
| 3.1 | A selective list of forest carbon accounting and assessment studies | 49 |
| 4.1 | Estimated growth and yield for the standard plus bedding and herbicide use regime (S+B+H) of slash pine plantation forests in the US South | 69 |
| 4.2 | Input cost and output price data of forest production in the US South | 70 |
| 4.3 | Estimated annual revenues, costs, profit, and timber and carbon outputs per acre for the standard plus bedding and herbicide use regime (S+B+H) | 71 |

4.4 Estimated carbon sequestration/storage outcomes (tons) under different scenarios for the S+B+H regime (carbon price is $20 per ton of $CO_2$) 73
5.1 China's cumulative C emissions (unit: petagrams) 83
5.2 China's C sequestration by forest ecosystems 84
5.3 China's carbon storage in harvested wood products (unit: petagrams) 85
5.4 A summary of the historical and projected carbon emissions and removals (Pg), and offset ratios (%) 86
6.1 Carbon emissions embodied in China's paper trade in 2014 103
7.1 Major categories of forest carbon finance 125
7.2 Carbon pricing alternatives 127

## Boxes

1.1 China's 1st Nationally Determined Contribution 11
1.2 China's forest data availability 15
1.3 The accounting framework and outcome for China 19
1.4 China's current forest conditions 20
6.1 European Union's carbon border tax—An excerpt from Böhringer et al. (2022) 99
7.1 Carbon tax in Costa Rica 117
7.2 Carbon cap-and-trade in California, United States 118
7.3 China's ecosystem restoration and conservation 124

# Acknowledgements

It was my great honor to have two excellent Forewords written for this book—one by Dr. John A. Parrotta and the other by Dr. William F. Hyde. Dr. Parrotta is a Program Leader of the USDA Forest Service on International Forest Science Issues and the current President of the International Union of Forest Research Organizations (IUFRO). Dr. Hyde has been a close friend of mine even since he lectured at Beijing Forestry University in the summer of 1987 when he was on the faculty of Duke University's School of Forest Resources. Their expert assessments and complementary remarks are sincerely appreciated.

I was also grateful for the strong evaluations and constructive comments provided by the three anonymous reviewers of my book proposal, which not only affirmed what I had proposed to write but also stimulated how I should get some of the topics addressed more coherently and adequately. Moreover, I was lucky to receive enthusiastic and generous book endorsements from Professors Shilong Piao of Peking University (China), Shashi Kant of the University of Toronto (Canada), Wan-Yu Liu of National Chung Hsing University (Taiwan), and Mónica Gabay of National University of San Martin (Argentina).

Many graduate students, in-service professionals, and visiting scholars, who took or audited the class I taught, *Forest Carbon Policy, Economics and Finance*, gave me beneficial feedback and kind encouragement. I was also fortunate to be able to collaborate and coauthor with many scholars, some of whom are talented and young, in both the USA and China on related topics. I am indebted to all of them. Further, I am pleased to acknowledge the financial support I received from the Office of Research and Innovation, AgBioResearch, and Department of Forestry of Michigan State University (MSU) in writing this book.

Moreover, I am gratified for the assistance and/or facilitation from many colleagues. In addition to David Skole and Richard Kobe of MSU, these colleagues from the North America and other parts of the world (than China) include: Qiang Ma of the Forestry Department of United Nations Food and Agriculture Organization; Bruce Manley and David Evasion of the University of Canterbury and Richard Yao of Sion, New Zealand; Peter Degeen of Technische Universität Dresden and Daniel Müller and Zhanli Sun of

Leibniz Institute for Agricultural Development in Transitional Economies, Germany; Sen Wang of the University of Toronto, Canada; Brent Sohngen of the Ohio State University, David Newman of the College of Environmental Science and Forestry, State University of New York, and David Wear of Resources for the Future, USA; and Nicola Caravaggio of Università degli Studi del Molise, Italy. Also included are colleagues from China: Wei Xie, Jintao Xu, and Jikun Huang of Peking University; Guomo Zhou and Yueqin Shen of Zhejiang Agricultural and Forestry University; Yahua Wang of Tsinghua University; Yali Wen of Beijing Forestry University; Shuifa Ke of the People's University of China; Hongqiang Yang of Nanjing Forestry University; Tianjun Liu and Han Zhang of Northwest Agricultural and Forestry University; Yongwu Dai of Fujian Agricultural and Forestry University; and De Lu of Asia-Pacific Network for Sustainable Forest Management and Rehabilitation and Can Liu of the National Forest Economics Research Center, State Forest and Grassland Administration.

Hannah Ferguson, Katie Stokes, Jonathan Merrett, Katie Hemmings, and their colleagues of Routledge provided high caliber coordinating and editing services in producing this book. Their courtesy, patience, and cooperation made my writing more efficient and enjoyable. I own them a heartfelt "Thank you."

Last but not least, I have been blessed to have the company of my wife, Xiu Du, and two daughters, Eva and Anna, in my journey of life and career. Their love, inspiration, and support have kept me going.

# Foreword

The career of the author, Professor Runsheng Yin, has been a remarkable contribution to the global material of forest economics and policy. It continues with this book which is an important addition to our still developing understanding of forests and global warming.

Most of us are fortunate to make one significant professional contribution. Professor Yin has made four. He was a key participant in the introduction of markets to Chinese forestry in the 1980s and papers from his dissertation on that topic have been fundamental to a plentiful subsequent economic literature on that topic, as well as to the global discussion of collective action and administrative transparency of which the still evolving Chinese policy has been an important part. His later research on commercial forest production in the US South includes work on a profit function for forestry, in my opinion the most important theoretical contribution to forest economics in the last century. The largest share of his academic career, since 1995, has featured lectures and seminars related to the alternatives of private or collective action in forestry. Some of this is a natural sequence from his dissertation. Some of it is entirely new. During this time, Professor Yin has directed the graduate or visiting scholar experience of more than 70 international students. And then, more recently, his attention has focused on the global role of forests in mitigating climate change—surely the crucial problem for forest policy in our modern time. This range of experience and continuing contribution is altogether unusual in one career and it is truly admirable in my opinion.

Regarding climate change, forests are an essential, even a pivotal, sector. As Professor Yin points out, forest sector activities alone could offer two-thirds of the cost-effective nature-based solutions necessary to hold global warming to less than the 2.0°C increase target of the 2015 Paris Agreement on Climate Change. Accordingly, forests must be an important component of any conversation about climate change and, similarly, climate change must be fundamental to any discussion of contemporary forest policy.

Various other analyses, including some substantial macroeconomic modelling (cited in this book), have examined the broad potential role of forests in mitigating climate change. Professor Yin's new book is the first, to my knowledge, to examine the internationally agreed policy requirements of the

Paris Agreement on Climate Change and their implications for the forest policy of the 194 national and international participants to the Agreement. This practical information is a necessary foundation for any further widespread on-the-ground progress on carbon sequestration and the mitigation of global change.

The Paris Agreement is easily understood, as Professor Yin outlines, but its application requires entirely new categories of forest measurement and reporting for every one of those 194 participants. Professor Yin describes these several categories—and the complexities of their application—in clear organization and with numerous national examples. His featured US and China examples are particularly meaningful as the US is the largest consumer of global carbon and China is both a very large producer, but an intermediate producer—which makes for new and different measurement problems for both China and the primary producers.

Professor Yin's book explains the incomplete, often difficult, sometimes competing, measurement and policy adjustments necessary to satisfy the Paris Agreement—and the differences they can make across regions within a country, between trading nations, and for the North-South economic development dichotomy. As such, it provides crucial focus and organization for further discussion of the implementation of global forest carbon policy and basic context for those who need to understand the association between the Paris requirements and their application, including both national policymakers and those who intend to extend further good analysis and assessment of this most crucial global issue of our day.

William F Hyde

# Foreword

Among the many good arguments for the conservation and sustainable management of the world's forests, the role of forests in the global carbon cycle and their importance for absorbing and storing atmospheric $CO_2$ is certainly a compelling one. While the 2015 Paris Agreement goal of limiting the increase in the global average temperature to well below 2°C above pre-industrial levels cannot be met without strengthening regulation of carbon-intensive industries and dramatically reducing emissions from the transport and energy sectors, among others—forests can nonetheless be part of the solution. Forests absorb as much as 29% of annual anthropogenic $CO_2$ emissions. Yet ongoing deforestation and forest degradation also result in significant emissions, equivalent to 10% of the annual anthropogenic $CO_2$ emissions.

Over the past 30 years, there have been a number of national, regional, and global efforts to enhance the role of forests as carbon sinks. These include the Clean Development Mechanism (CDM) of the Kyoto Protocol and more recent REDD and REDD+ initiatives[1] developed in the context of the UN Framework Convention on Climate Change (UNFCCC), as well as other initiatives related to forest landscape restoration. Experience has shown that such efforts can yield positive results from a carbon perspective where they also provide significant and lasting benefits to the people most directly affected by the changes in forest and land use promoted by these initiatives. However, experience has also shown that failure to address the direct and indirect drivers of deforestation and forest degradation operating at different spatial scales can doom or severely limit the permanence of carbon gains from such efforts. Among the most important of these drivers are issues related to land rights, tenure, lack of intersectoral policy coordination, and a variety of governance issues that influence land-use and investment decisions at local, subnational, national and global scales.

Beyond the difficulties in designing and implementing effective actions to halt and reverse forest loss and degradation, there are also significant challenges in measuring progress towards meeting national and collective commitments for climate action under the Paris Agreement and associated forest-related programs initiatives, including results-based-payment schemes under REDD+. Although advances in the use of remote sensing have greatly facilitated assessments of forest area change, degradation remains much more difficult to monitor. In the case of REDD+ actions, accurate measurement, reporting and verification of carbon outcomes are limited by discrepancies between different datasets, lack of country-specific data, inadequate reporting on estimate uncertainty, insufficient resolution of satellite imagery to monitor forest degradation, lack of inclusion of other carbon pools such as deadwood or soil carbon, and the uncertainty surrounding the impact that climate change will have on forests and their carbon sink function.

This book focuses on a number of critical issues encountered in international forest carbon accounting and related policy assessment, including the differential potentials of carbon sequestration by forest ecosystems and storage by harvested wood products, the joint production of timber and forest carbon offsetting, and results-based payments, among others. It draws on experiences throughout the world, with a special emphasis on China and the United States.

This work opens with an overview of the Paris Agreement and its implications for the forest sector (in Chapter 1), and an examination of jurisdictional and other approaches to governing forest climate solutions (in Chapter 2). Chapter 3 provides a review and synthesis of the different perspectives of and approaches to forest carbon accounting and economic assessment, with case studies from North America and Europe, followed by a discussion of alternative approaches to forest carbon accounting at the micro level based on the pine plantation forests in the US South (in chapter 4), and an assessment of the forest sector's potential role in China's decarbonizing drive (in Chapter 5). Chapter 6 examines the issue of carbon leakage through an evaluation of China's wood products trade, highlighting difference between production- vs. consumption-based carbon accounting approaches. Chapter 7 provides a review of carbon finance and forest sector climate funding in particular, while Chapter 8 examines issues related to the identification and analysis of risk and uncertainty, and discounting entailed in coping with climate change. Concepts and techniques for impact evaluation of forest carbon projects or programs are explored in Chapter 9, with case studies from Africa, Southeast Asia, and Latin America. The book concludes (in Chapter 10) with a consideration of the implications and lessons learned from the preceding chapters, consideration of knowledge gaps and future directions for work in this field.

Professor Yin is to be commended for his comprehensive scholarship and contributions to our understanding of the rapidly-changing fields of forest carbon accounting, economics and finance and its implications for policymakers. This work will undoubtedly advance our collective efforts to optimize the role of forests and the forest sector in climate change mitigation and adaptation.

John A. Parrotta

**Note**

1 REDD+: Reducing emissions from deforestation and forest degradation, conservation of forest carbon stocks, sustainable management of forests, and enhancement of forest carbon stocks.

# Introduction

This book is about the forestry component of the 2015 Paris Agreement on climate change—probably the most important forest policy of our time and certainly the first commonly agreed upon global forest policy. The Agreement brought all Parties to the United Nations Framework Convention on Climate Change (UNFCCC) together for a common purpose: "holding the increase in the global average temperature to well below 2°C ... and pursuing efforts to limit the temperature increase to 1.5°C above pre-industrial levels" (Article 2a).

Forests play a crucial role in mitigating climate change. Globally, land-based activities, including agriculture, forestry, wetlands, and bioenergy, could sustainably contribute about 30% (15 billion tons) of the $CO_2$ equivalent per year required to limit the increase in the temperature to 1.5°C above pre-industrial levels by year 2050 (Roe et al. 2019). Forest sector activities alone could offer two-thirds of the cost-effective nature-based solutions necessary to hold warming to less than a 2.0°C increase (Griscom et al. 2017). Applications of forest policy to achieve these targets are the subject matter of this book.

The Paris Agreement is predicated on nationally determined contributions (NDCs). It offered each of the 194 signatory Parties (193 national governments and the European Union) the opportunity to devise its own preferred strategy for satisfying the global mitigation and adaptation targets as well as the discretion to choose its own indicators for measuring progress toward those targets. It is thus not surprising that the many national initiatives and follow-up attempts of ambition raising have resulted in a wide variety of climate and forest strategies (IPCC 2020, FAO 2016).

The objective in this book is to address the major issues encountered in the variety of these strategies, beginning with a perspective on and elaboration of the Paris Agreement itself (Chapters 1 and 2) before empirically assessing mitigation efforts at different levels of aggregation (Chapters 3–5). Chapters 6–8 examine three broader issues that extend beyond national boundaries or sovereign national goals: international trade, funding and finance, and risk, uncertainty, and social discounting. These chapters also include examples from many parts of the world. Chapter 9 presents a review and synthesis of impact assessments of forest carbon policies, focusing on the popular projects and programs of REDD+ (i.e., reducing emissions from deforestation and

DOI: 10.4324/9781003436652-1

## 2  Introduction

forest degradation, conservation of forest carbon stocks, sustainable management of forests, and enhancement of forest carbon stocks). The final chapter provides summary observations and recommendations for further inquiry and revised subnational, national, or global policies.

Five issues crucial to deliberating the forest role in the mitigation of climate change run through the discussion:

a   jurisdictional approaches to climate governance,
b   the differentiated amounts of carbon sequestered in various forest ecosystems, as reflected in biomass, both above/below ground and in soil,
c   the effect of the joint production of timber and carbon storage,
d   the different measures of carbon stored in harvested wood products and the differential duration of this storage, and
e   the accounting and crediting requirements of additionality, permanence, and avoidance of leakage for forest sector climate solutions.

The first, jurisdictional orientation, has its own chapter but continues as a thread in following chapters. The remaining four appear repeatedly in each of the other chapters. It will be useful to grasp the meaning of each as it first arises, for it will continue to be important in further discussion.

These five issues are each complex and taken together make carbon accounting and assessment even more complex for some larger countries that host multiple forest ecosystems and produce a range of goods and services. Each creates further new complications for the comparison of forest carbon strategies across participating countries and for the aggregation of national impacts necessary to assess a global outcome of the carbon account. Discussion of the means and measures to successfully address these issues will be an important task for the final chapter and for the future success of policies.

In short, this book attempts to tackle a set of important issues inevitably faced in exploring the forest sector's role of climate change mitigation by providing the policy background and research development, introducing (or building) necessary modelling approaches, featuring illustrative examples from different parts of the world, and summarizing the outstanding difficult challenges of policy and measurement. The overall objective is to develop a coherent knowledge base regarding the numerous processes and their varied principles and practices for the reduction of greenhouse gas emissions and the enhanced removal (through sequestration and storage) of these emissions. Such a coherent knowledge base is essential for policymakers, analysts, and practitioners who strive to grasp the breadth of this complex topic.

### Background on Global Climate Agreement

Following the 1992 UNFCCC, the Kyoto Protocol of 1997 was the first institutionalized agreement on global climate change. Participants in the Protocol were UN member nations, known as Parties, who have continued to this day to meet annually under UN auspices to discuss climate change. The Protocol required industrialized countries to reduce their greenhouse gas emissions, but it did not impose binding obligations on any developing countries.

This dichotomous distinction between the developed and developing countries in the Kyoto Protocol made progress on climate change impossible, because growth in emissions since the Protocol came into force in 2005 is entirely in the large developing countries, such as China, India, Brazil, Korea, South Africa, Mexico, and Indonesia (Stavins 2015). Countries covered by the Protocol (i.e., Europe and New Zealand) accounted for no more than 14% of global gross emissions (and 0% of global emissions growth). In contrast, 186 members of the UNFCCC submitted their intended NDCs by the end of the Paris talks, representing some 96% of global emissions.

Also, only afforestation and reforestation projects were included as part of the designated mitigating efforts under the Kyoto Protocol's Clean Development Mechanism. Later, in 2007, the Bali Action Plan was adopted by the UNFCCC to reduce greenhouse gas emissions from deforestation and forest degradation (i.e., REDD). In 2010, the Cancun decision on REDD+ was adopted to expand the role of forests in mitigating climate change to include conservation of forest carbon stocks, sustainable management of forests, and enhancement of forest carbon stocks (Parrotta et al. 2022). REDD+ was conceived as a voluntary, nationally driven mechanism for high-income countries (as listed in Annex 1 of the UNFCCC) to pay low- and middle-income countries (non-Annex 1 Parties to the UNFCCC) for the reductions in forest emissions. Furthermore, the NDCs have themselves been evolving. For instance, 76% of the first generation NDCs included land use, land use change, and forestry (LULUCF), whereas 90% of the second-generation NDCs, those submitted to the UNFCCC by December 2020, include LULUCF.

Therefore, the empirical knowledge before the Paris Agreement was limited to the project experiences of Kyoto and REDD+, but much of this is no longer adequate. Moreover, coupled with the changed rules of "the carbon game," the diversity of national programs leaves us with confusion and controversy regarding the governance of climate actions, carbon offsetting, accounting and credit taking, and assessed REDD+ results, among other things. In addition, insufficient efforts have been made to examine the more complete set of issues identified above or the issues of adjustments for international trade or means of funding and finance discussed in Chapters 6 and 7. Thorough comprehension of every one of these issues is necessary for a reasoned understanding of the forest sector's contribution to climate change mitigation and for the more difficult global assessment required by the Paris Agreement.

Therefore, the diversity of existing experience and the difficult but urgent global aggregation are what make the material of this book important. An organized assessment of the multiple current strategies, targets, and practices for the reduction of greenhouse gas emissions and the enhanced removal of these emissions is the objective and it is essential for understanding the forest role in climate change.

4  *Introduction*

**Organization**

The ten chapters that follow begin with a review of the relevant provisions of the Paris Agreement and their implications for the forest sector in Chapter 1. The central provision in the Agreement is that each Party determines its own strategies and has the discretion to assign its own indicators enumerating progress toward both its national goals and targets in relation to the global goals and targets of the Agreement. Any Party may have multiple goals, but their key components should include measures of the greenhouse gas emissions and removals designated to satisfy the targets included in its goals. The Agreement requires each Party to assess its progress in this respect biannually, and each Party's assessment, when adjusted for their complex differences, becomes input for the accumulative global assessment.

The forest sector actions, specified in Article 5.2, are most relevant for our interest. They include the obvious issues of deforestation and forest degradation, conservation, and forest management—and it is worth considering how these requirements of Article 5.2 differ from REDD+. Accordingly, a more systematic and consistent forest carbon accounting is called for, covering carbon sequestered and stored in forest biomass (above and below ground), soil, and harvested wood products. Further yet, Article 4.1 requires an accounting of emission reduction and removal resulting from human actions. We'll see in later chapters how this imposes a new challenge to national forest carbon accounting. This first chapter concludes with comments on the differences of jurisdictional and other approaches to climate governance.

Chapter 2 follows with more detail on the different approaches to climate governance. Under Kyoto and REDD+, carbon accounting rarely went beyond the project. Independent carbon projects might be organized within a smaller district of a country or within a single government agency, and their impacts were not accumulated. Following the Paris Agreement, all projects, regardless of the regional governing body or national agency, become part of the national targets. This reorientation is certainly advantageous and beneficial in terms of accountability, capability, and reliability of achieving the pledged NDCs, among other things. But it also introduces problems of accounting across jurisdictions within a single country, including questions of additionality that arise when the action of one region or sector offsets the action of another, which furthers issues of country-wide equitable investment and benefit sharing across subnational jurisdictions, issues that different regions or different agencies within a country have seldom confronted previously.

Within the forestry component of all of this, forest measurement has always been the responsibility of national forest agencies, and the focus of these agencies has always been on their own measured inventories of standing timber and forest land. Their inventories must now extend to the national requirement to measure carbon both above and below ground, and this new responsibility introduces the forest agencies to a much broader client base

Introduction    5

than that of their previous experience. Chapters 1 and 2 highlight these differences for forest measurement and discuss the new requirements for forest agency carbon measurement. Overall, these two chapters are foundational to any discussion of forest carbon policy following the Paris Agreement.

Chapter 3 overviews the features that make accounting and assessment within the agriculture and forest sector unique, then discusses the two fundamental alternative approaches to these analytic endeavors, as well as their data requirements. The first is a direct approach based on the traditional forest inventory collected by national forestry agencies but extrapolated in two different directions, first to add elements of carbon not included in pervious above-ground measures of the forest stock and then extrapolate in a different manner to add carbon stored in harvested forest products.

The second, altogether different, approach is an indirect approach based on market modelling of forest product supply and demand. For this latter approach, the same standing forest inventory used in the direct approach is a shift variable for supply and, once more, forest biomass must be extrapolated from the inventory variable. Soil carbon, as yet, has not been included in calculations using this approach. The indirect method, to this day, has rarely been applied beyond scholarly explorations in large global, continental models. In addition to spotlighting an assessment of global forestry development since 1990 that features the indirect approach, the chapter showcases national applications of the direct approach: Canada, Europe (essentially the European Union), and Russia. Comparison of the latter two cases—Europe and Russia—also raises the question of the limits to, or the saturation level for, carbon storage in mature forests.

Chapters 4 and 5 review the empirics of forest carbon modelling, first at the local and micro scale relevant to individual forest landowners (Chapter 4) and then at the national level (Chapter 5). Each perspective has been a focus within the policy discussion and also the literature on the forest role in global climate change, but neither solves all the problems identified with forest carbon accounting. Following an introduces to the nature-based solutions to the climate crisis, Chapter 4 confronts the current analytic approach, essentially a single stand model with outputs of carbon and timber jointly. Some of the weaknesses of this model can be corrected with a profit function formulation that includes forest stands of different ages and different locations and also add compensation for incorporating the costs of carbon sequestration and storage. A detailed empirical application for southern pine in the United States illustrates this alternative model, its results, and their sensitivity to various forest stand and market conditions.

Chapter 5 elucidates the bigger picture of national modelling, with China as the example. China is important as a large country, global carbon emitter, and a signatory to the Paris Agreement. China is further interesting, as both its standing forest volume and its forest area are growing and thereby increasing their sequestration of carbon. The assessment in this case builds on the combination of three large data intensive national models, one each for emissions, sequestration, and the storage of harvested wood products. Alternative scenarios for each, projected over time and aggregated, show a range of the potential for forests to satisfy China's targets in offsetting carbon emissions. The potential effect of the forests can be impressive, but achieving its full potential will require important modifications of current policy. These are not unlike forest policy changes that will be required in some other countries. The conclusion to the chapter discusses these policy modifications.

Chapters 6–8 discuss three broader issues extending beyond national boundaries that affect each country's carbon accounting and assessment in its own individual way. The first is international trade, and the discussion of trade includes a variety of problems associated with adjustments for carbon leakage and border taxes. Leakage occurs when carbon accounts are measured in production, as is usual, but that measure in production may not be the same as that in consumption. This is the common situation for the many developed countries that import more forest products than they export. It's the opposite for many developing countries. As global trade expands, this difference has increased—and it promises to continue to do so. Two recent studies, one of global land use and the other of China's paper industry, together with my own assessment of carbon emissions embodied in China's forest products trade, show the prevalence and magnitude of the problem. The chapter continues, explaining how border taxes, depending on their form, could either create an even greater negative bias for those resource-exporting developing countries, or they could correct the leakage problem. However, fair and accurate border taxes would require international taxation authority—something that not even the European Union has to date.

The global investment necessary to limit the catastrophic effects of climate change has been estimated in the neighborhood of US$9.2 trillion annually, with a gap of at least US$3.5 trillion over today's expenditure. Chapter 7 reviews and summarizes the literature and practice regarding the funding of forest sector climate solutions. It begins with discussion of the principles of carbon pricing and continues with a deliberation of recent developments of both private and public roles for funding forest sector solutions. Different initiatives, not surprisingly, have different means for their funding and different relative costs. The chapter closes with a review of the best-known forest carbon funding programs: REDD+ and market-based programs, both with high transactions costs, additionality and leakage, and jurisdictional problems. It also raises the question: Could pooling or aggregation across international boundaries be a solution?

Chapter 8 discusses the problems of risk, uncertainty, and social discounting for the stream of effects of national programs to affect carbon. The long-time horizon and the possibility of major, non-marginal effects make these topics crucial to carbon accounting and assessment. The global and non-private nature of the problem raises the question of social discounting. The chapter discusses the Ramsey approach used in the Stern Review (2006) and the assessment of its three terms: time preference, the marginal elasticity of utility, and the per capita growth rate of consumption; it then introduces the intergenerational question of the impact of a declining discount rate. A final section of the chapter overviews Nordhaus's critique of Stern's discounting and highlights the prospective of long-term financing under a term structure of discount rates.

Chapter 9 explores the concepts and techniques of impact evaluation for forest carbon projects and programs, with examples from Africa, Latin America, and Southeast Asia. The chapter reviews the concepts and methods of impact evaluation that might be applied, then discusses sampling and data, including the crucial issue of an appropriate baseline from which to conduct any measurement over the duration of a project or program. This review is important for identifying the approaches and problems that arise when evaluating individual projects and programs within a country's carbon strategies as required by the Paris Agreement, especially those regarding the identification of the business as usual, or counterfactual, scenario at the highly disaggregate level.

Finally, Chapter 10 considers the policy and research implications to be drawn from the preceding chapters, the further work to be undertaken, and the questions remaining to be answered in the future. It closes with several takeaway messages and one last appeal—a lot more must be done in advancing the research agenda that I have laid out in this book—to realize the high expectations that the global forest sector should deliver in climate change mitigation as well as adaptation.

In sum, the fundamental problem posed by climate change is that of a public good or externality; in this case, a negative externality or public "bad" in the form of greenhouse gas emissions. Climate change is a particularly thorny externality because it is global—its impacts are indivisibly spread around the entire world—and it resists the control of both markets and national governments (Nordhaus 2019). Therefore, governance is a central issue because effective management requires the concerted action of many countries (Stern 2006). It is only by designing, implementing, and enforcing cooperative multinational policies that nations can ensure effective climate-change policies.

This book aspires to place the technical and policy issues regarding the forest sector component of the global externality in appropriate context and to build useful perspectives, frameworks, and methods for addressing them successfully in the future. It intends to identify and fill some of the salient voids of the literature and to contribute a more comprehensive understanding of global forest carbon policy, economics, and finance.

## References

Food and Agriculture Organization of the United Nations (FAO). 2016. *Forestry for a Low-Carbon Future: Integrating Forests and Wood Products in Climate Change Strategies* (FAO Forestry Paper 177). Rome, Italy.

Fuss, S., Lamb, W.F., Callaghan, M.W., Hilaire, J., et al. 2018. Negative emissions—part 2: Costs, potentials, and side effects. *Environmental Research Letters* 13 (6); https://doi.org/10.1088/1748-9326/aabf9f.

Griscom, B.W., Adams, J., Ellis, P.W., Houghton, R.A., Lomax, G., Miteva, D.A., Schlesinger, W.H., Shoch, D. et al. 2017. Natural climate solutions. *PNAS* 114 (44): 11645–11650.

International Panel on Climate Change (IPCC). 2020. *Climate Change and Land: Summary for Policymakers*. Available at: www.ipcc.ch/srccl/.

International Panel on Climate Change (IPCC). 2022. *Climate Change 2022: Impacts, Adaptation and Vulnerability*. Working Group II Contribution to the Sixth Assessment Report.

Nordhaus, W. 2019. Climate Change: The ultimate challenge for economics. *American Economic Review* 109: 1991–2004. Available at: https://doi.org/10.1257/aer.109.6.1991.

Parrotta, J., Mansourian, S., Wildburger, C., Grima, N. (eds). 2022. *Forests, Climate, Biodiversity and People: Assessing a Decade of REDD+*. IUFRO World Series Volume 40. Vienna.

Roe, S., Streck, C., Obersteiner, M., Frank, S., Griscom, B., Drouet, L., et al. 2019. Contribution of the land sector to a 1.5 °C world. *Nature Climate Change* 9: 817–828 doi:10.1038/s41558-019-0591-9.

Stavins, R. 2015. *An Initial Assessment of the Paris Agreement*. Harvard Project on Climate Agreements.

Stern, N. 2006. *The Economics of Climate Change: The Stern Review*. Cambridge University Press. Available at: http://mudancasclimaticas.cptec.inpe.br/~rmclima/pdfs/destaques/sternreview_report_complete.pdf.

United Nations Framework Convention on Climate Change (UNFCCC). 2020. *Reference Manual for the Enhanced Transparency Framework under the Paris Agreement*. Bonn, Germany.

UN REDD Programme. 2022. *Linking REDD+, the Paris Agreement, Nationally Determined Contributions and the Sustainable Development Goals: Realizing the Potential of Forests for NDC Enhancement and Implementation*. Available at: www.un-redd.org/sites/default/files/2022-03/NDC%20Final.pdf.

# 1 A primer on forest carbon policy and economics under the Paris Agreement

**Introduction**

Forest sector climate solutions can be both significant and cost- and time-effective (Roe et al. 2019, Griscom et al. 2017), as I have already noted in the Introduction. Given that, it is fundamental for forest economic researchers and policy makers to build a clear and consistent understanding of the processes, principles, and rules pertinent to climate change mitigation and adaptation under the Paris Agreement, which constitutes the objective of this chapter. Doing so will enable them to tackle the challenges concerning carbon accounting, climate governance, policy assessment, impact evaluation, and other issues more effectively.

I will pursue that objective by highlighting the core elements of the Paris Agreement and deliberating on their implications for forest sector actions. My exploration will take into account the unique characteristics of forest ecosystems (FESs) that make it challenging to handle these actions properly. Whenever necessary, I will also use case studies or examples to illustrate my viewpoint. Specifically, I will focus on the following interrelated issues: the makeup and information needs of the nationally determined contributions (NDCs), the processes of progress tracking and stock taking, jurisdictional approaches to climate governance, the determination of anthropogenic greenhouse gas (GHG) removals, the carbon accounting principles and rules, and the required forest sector data and gaps in them. Finally, I will conclude with a few remarks concerning how to capture various carbon pools of FESs adequately and how to take advantage of their role in mitigating climate change by strengthening conservation and accelerating growth.

Several of the issues have been discussed and even debated in the past. To name a few, Sedjo and Sohngen (2012), Gren and Zeleke (2016), and Thamo and Pannell (2016) examined challenges of forest carbon accounting with respect to additionality, permanence, and/or leakage avoidance; Fischer et al. (2016) reviewed the studies of REDD+ projects,[1] while Chagas et al. (2020) and Galik et al. (2016) elucidated their governance; Boyd et al. (2018) analyzed the jurisdictional approach to land-based mitigation and adaptation actions; and Juutinen et al. (2018) and van der Gaast et al. (2018) assessed the

DOI: 10.4324/9781003436652-2

10  *A primer on forest carbon policy and economics*

feasibility and potential of forest-based carbon sequestration and compensation schemes. Nonetheless, certain processes, principles, or rules have not been interpreted or articulated clearly, resulting in controversy and confusion.

It should be acknowledged that this chapter is based on a paper my colleagues and I published under the same title (Wang et al. 2021). In preparing that paper, we drew information from key documents of the leading international organizations, including: *Reference Manual for the Enhanced Transparency Framework under the Paris Agreement* (UNFCCC 2020), *Forestry for a Low-Carbon Future: Integrating Forests and Wood Products in Climate Change Strategies* (FAO 2016), and *2006 Guidelines for National Greenhouse Gas Inventories* (IPCC 2006). I have made numerous changes to that paper when writing this chapter.

## Understanding the Paris Agreement

My interpretation and elaboration of the relevant processes, principles, and rules of the Paris Agreement are structured into the following two sub-sections: one on the general stipulations and requirements, including NDC targets and indicators, progress tracking and stock taking, and jurisdictional governance; the other on forest-sector stipulations and requirements, covering specific actions, reference levels, anthropogenic testing, and carbon accounting principles.

### *General stipulations and requirements*

#### *Targets and indicators*

The Paris Agreement has offered Parties not only the opportunity to come up with their own NDCs but also the discretion over what indicators they may use to enumerate the goals in their NDCs. So, it is useful to begin with a synopsis of the NDCs' makeup and the related information requirements for their execution and communication. Overall, included in the goals of the NDCs are: economy-wide and/or sector-specific absolute emission reduction or limitation target relative to a base year, emission reduction target below a 'business as usual' level, intensity target, and/or peaking and net-zero targets (UNFCCC 2020). As detailed later, forest sector actions constitute efforts of emission reduction through efforts of controlling deforestation and forest degradation, and emission removal through afforestation, reforestation, and forest management, in addition to carbon storage in harvested wood products (HWPs).[2]

Parties may have multiple types of goals in their NDCs, as shown in China's first NDC (see Box 1.1). In preparing and communicating an NDC, the key components of information should comprise GHG emissions and removals for those chosen categories of targets the NDC covers. Thus, a Party begins by articulating its concrete actions of emission reduction and removal and selecting indicators for the actions; once the indicators are selected, data

*A primer on forest carbon policy and economics* 11

should be gathered on each of them, including reference information, any updated statistics from previous years, and the latest available values for them. Just like an NDC can be disaggregated into sector-specific actions and targets, it can also be decomposed into subnational actions and targets. Sectoral and subnational actions and targets are normally nested within the NDC (Chagas et al. 2020, Wunder et al. 2020).

> **Box 1.1 China's 1st Nationally Determined Contribution**
>
> In its first NDC submitted to the UNFCCC in 2015, China expressed its intentions to:
>
> 1 achieve peak $CO_2$ emissions around 2030 and make its best efforts to peak earlier;
> 2 lower $CO_2$ emissions per unit of GDP by 60–65% from the 2005 level;
> 3 increase the share of non-fossil fuels in primary energy consumption to around 20%; and
> 4 increase the forest stock volume by 4.5 billion cubic meters ($m^3$) from the 2005 level.
>
> (NDRC 2015)

To ensure clarity and transparency, it is crucial for a Party to gather information on the sources of data used in quantifying reference point(s) and indicators, and under what circumstances it may update the values of those reference indicators. Also, there must be careful documentation of the assumptions and methodologies used for accounting for GHG emissions and removals. Parties are required to select their methodologies and metrics in accordance with the 2006 IPCC Guidelines.

*Progress tracking and stock taking*

The Paris Agreement mandates that the Conference of Parties (COP) "shall periodically take stock of the implementation of this Agreement to assess the collective progress towards achieving the purpose of this Agreement and its long-term goals" (Article 14.1). Article 4.13 further stipulates that

> Each Party shall regularly provide the following information: (a) A national inventory report of anthropogenic emissions by sources and removals by sinks of greenhouse gases, prepared using good practice methodologies accepted by the IPCC and agreed upon by the COP ... and (b) Information necessary to track progress made in implementing and achieving its NDC.

As such, the processes for tracking and updating national-level efforts have been well set to evaluate whether countries are meeting their goals—national progress tracking—and whether the collective sum of individual NDCs is on track to meet the overall purpose of the Paris Agreement—global stock taking (UNFCCC 2020, FAO 2016).

In other words, national progress tracking and global stock taking are the main international processes for implementing the NDCs. While these processes are related, there are differences between them. Before delineating these differences, though, it is beneficial to introduce the Enhanced Transparency Framework (ETF) defined in the PA: "Parties shall promote environmental integrity, transparency, accuracy, completeness, comparability and consistency, and ensure the avoidance of double counting, in accordance with guidance adopted by the Conference of the Parties" (Article 4.13). Transparency refers to the reporting of information by a Party in its biannual transparency report (BTR), including information on the GHG inventory, the accounting approach(es) selected, and the indicators used for tracking progress and, ultimately, to demonstrate whether they have achieved their NDCs.

In addition to national measurement as well as monitoring, reporting, and verification (MRV), international review of that information will be carried out by technical committees and/or through procedures organized by the COP. Information gained from reporting and review under the ETF acts as an input to the global stock taking, which will be undertaken every five years (the first due in 2023). The ETF should in turn encourage higher ambition by making more readily available information on progress made by Parties achieving their NDCs (UNFCCC 2020). While the targets specified in an NDC are not legally binding, Parties, especially the major ones, are strongly urged to not only fulfill their pledges but also raise them if necessary.

For instance, the UNFCCC (2015) and other relevant organizations, upon assessing the initial NDCs announced by Parties, found that they would be unlikely to achieve the overall goal of controlling the temperature rise within 2°C in this century. The UNFCCC then urged the major Parties to be more ambitious. In response to this call, numerous countries have upped their ante by announcing the realization of carbon neutrality no later than the mid-century and/or raising some of their climate targets. As an example, China pronounced its goal to achieve net-zero emissions before 2060 and increase in forest stock volume by 6 billion m$^3$ by 2030 while pursuing other targets, including international assistance, more aggressively (Energy Foundation China 2020). Likewise, in its Enhanced NDC submitted to the UNFCCC Secretariat on September 23, 2022, Indonesia raised its emission reduction target from 29% in the initial NDC to 31.89% unconditionally and from 41% to 43.20% conditionally and unconditionally combined by 2030 (Republic of Indonesia 2022).

*Jurisdictional governance*

The NDC architecture features a jurisdictional approach to governing climate mitigation solutions, which has some distinct advantages. First, it provides a greater likelihood for Parties to carry out necessary actions and assume their responsibilities. A jurisdiction, be it a state, district, or municipality, can be held accountable in its performance against what it has pledged. In comparison, a non-jurisdictional entity, for example, a local non-governmental organization, may not have the trust, authority, or capability to deliver its pledge over time. A jurisdictional approach also makes it more practical for the carbon accounting principles and rules to be materialized. With a few stands of trees, for instance, a smallholder may find it hard to maintain his/her forest carbon stock permanent, whereas a county or state has little problem in so doing. While allowing logging and other commercial activities to continue, it can easily reach an additionality to its carbon stock and maintain it over time.

Similarly, leakage induced by trade flows across borders or land use changes can be monitored and determined without causing any concern about the final accounting outcome. In the next sub-section, I will elaborate on the principles of additionality, permanence, and avoidance of leakage in forest carbon accounting. Furthermore, it has been argued that within the forest sector, the jurisdictional approach is a recognition of the necessary and vital role that governments at multiple levels must play in any realistic effort to protect forests, reduce emissions, and enhance removals (Boyd et al. 2018). Governmental and non-governmental organizations should thus work within a given jurisdiction and complement the authority in delivering its pledged contribution.

A jurisdictional approach to governing forest climate solutions also offers important opportunities for experimentation and policy innovation, as well as controlling the transaction costs, including partnerships with supply chain actors and indigenous and traditional communities. Challenges to such an approach include political turnover, limited public sector capacity, fair benefit sharing, and lack of broader support and incentives (Boyd et al. 2018, Phelps et al. 2010)—all of which can hamper the kind of long-term, sustained attention these initiatives require to succeed. Recognizing its great relevance, however, Parties have adopted the NDC as a bedrock of the Paris Agreement. Thus, it is crucial for our policy dialogue to adequately reflect this fundamental approach, and forest carbon projects should be nested within a jurisdictional program (Seymour and Busch 2016).

### Forest-sector stipulations and requirements

#### Specific actions

Article 5.2 of the Paris Agreement specifies

> policy approaches and positive incentives for activities relating to reducing emissions from deforestation and forest degradation, and the role of conservation, sustainable management of forests and enhancement of forest carbon stocks in developing countries; and alternative policy approaches, such as joint mitigation and adaptation approaches for the integral and sustainable management of forests.

This clearly indicates the inclusive scope of forest sector actions. Someone may immediately realize that the first part of this statement is almost exactly those actions under REDD+, as originally advocated by the forest science and policy community (e.g., Nepstad et al. 2013, Seymour and Busch 2016).

Now, however, they can be carried out by not just tropical developing countries where there have been deforestation and forest degradation but also many others, such as Canada, China, and India, that have included forest sector actions in their NDCs. The latter part of the above statement pertains to developed country Parties in their international implementation of forest sector actions. Notably, actions taken within developed countries involve mostly domestic financing, whereas their international cooperation with developing countries undergoing deforestation and/or forest degradation is predicated on the former's payments for the outcomes of latter's REDD+ actions (FAO 2016, Seymour and Busch 2016). In general, Parties cooperate internationally in mitigation and adaptation by means of finance, technology development and transfer, and capacity-building initiatives (UNFCCC 2020).

As forest carbon stock is made up of carbon sequestered by FESs and stored in HWPs, the needed information for forest carbon accounting includes inventories of forest biomass (above and below ground), forest soil, and in-service wood products. Generally speaking, forest biomass has not been historically measured as part of a typical national forest inventory (NFI) system, which was and still is focused on the stock volume of regular forests (Liu and Yin 2012, Zeng et al. 2016). But forest biomass volume can be straightforwardly derived from the stock volume by establishing a relationship between them using the biomass expansion factor, and then carbon stock can be estimated from biomass stock using the appropriate carbon conversion factor (Fang et al. 2001, Piao et al. 2009). Otherwise, more sophisticated ecosystem models, based on in-situ and/or geospatial observations, can be deployed to quantify fluxes of various biomass and even soil carbon pools (Luyssaert et al. 2010).

Meanwhile, carbon stored in HWPs can be estimated using the FORSTAT database provided by the FAO (Johnston and Radeloff 2019, Geng et al. 2019) or another source in conjunction with the method(s) recommended by IPCC (2006, 2019). Evidence suggests that in many countries, carbon stored in HWPs is limited, but it is sizable in some large countries like Canada, China, and the US (Johnston and Radeloff 2019). Overall, it is a relatively small portion of the forest sector carbon storage (IPCC 2006).

The accounting and assessment challenges often lie in forest soil, which has rarely been systematically measured, and trees outside of forests (TOFs) and urban trees and forests, which have often been inadequately covered by the NFI (FAO 2020a).[3] As such, it is hard even in a developed country to have comprehensive inventory data for FESs (FAO 2020b). Without all the required inventories, it would be unlikely for a Party to conduct a comprehensive forest carbon accounting or policy assessment. As a result, offset credit taking and results-based payments can only be partial. For instance, a

Party may be able to account for the forest sector actions based on forest biomass pools and HWP storage, with the forest soil, TOF, and urban tree/forest pools being left out. Box 1.2 highlight China's FES data availability.

> **Box 1.2 China's forest data availability**
>
> Since its inception in the early 1970s, China has carried out 9 successive NFIs at an interval of 5 years, based on permanent sampling plots (~415,000). Later, those permanent plots were supplemented with temporary ones selected by means of geospatial tools. From the reports generated, we can identify the 2005 reference levels and estimate the forest area and stock volume changes over time (NFGA 2019). But China's NFI has kept a merchantable-timber orientation, indicating that the measured stock volume does not cover the totality of above-ground biomass, let alone the below-ground counterpart. Due to the efforts of developing the biological expansion and C conversion factors for various forest types, carbon sequestered in the total biomass can now be derived (Fang et al. 2001). So, the additional carbon stock can be determined for a given period of time. On the other hand, forest soil carbon has not been part of the NFI measurement (Piao et al. 2009). While soil is a large carbon pool of the FESs, its magnitude is more uncertain and less reliable.

*Anthropogenic testing*

The NDCs are designed "to achieve a balance between anthropogenic emissions by sources and removals by sinks of GHGs" (Article 4.1). In principle, any contribution a Party makes must come from human actions, with the exclusion of effects of natural causes. That means that the outcome of a mitigation action must pass the 'anthropogenic' test (Grassi et al. 2018). Here, I draw on two linked articles in a recent issue of *Nature Climate Change* (Grassi et al. 2021, Ogle and Kurz 2021) to illustrate this point.

Estimating human-caused $CO_2$ emissions is straightforward regarding the combustion of fossil fuels because the fossil fuels are extracted through mining or exploration. Therefore, all emissions are deemed anthropogenic. On the other hand, attributing emissions and removals to human activity in land-based systems is more complicated because even in the absence of human activity, there are natural effects associated with removals of $CO_2$ by land through photosynthesis by plants, or C emissions through decomposition of detritus and soil organic matter and by wildfires (Ogle and Kurz 2021). Direct human activities impact $CO_2$ emissions and removals from land, such as deforestation with conversion of forests to non-forest uses and, conversely, with forest re-growth. Much more difficult to quantify are land-sector emissions and removals from indirect human activities resulting from climate change, increases in atmospheric $CO_2$ concentrations, and nitrogen deposition (Ogle and Kurz 2021).

Climate scientists often use integrated assessment models (IAMs) to gauge progress toward limiting global warming by accounting for human activities that impact $CO_2$ fluxes directly, such as deforestation, timber harvesting, and other land-management practices. Grassi et al. (2021) show that $CO_2$ emissions and removals estimated by IAMs are inconsistent with estimates that governments compile in their national GHG inventories to inform their mitigation targets. Estimated to be over 5 $GtCO_2$ $yr^{-1}$, this discrepancy is mostly due to conceptual differences in the characterization of human-caused emissions and removals of $CO_2$ from land-based systems in IAMs and national GHG inventories. Consequently, a large fraction of land emissions and removals resulting from indirect human activities are considered 'natural' in IAMs but 'human-caused' in inventories.

By combining these indirect emissions and removals from dynamic global vegetation models with the direct emissions from the IAMs, Grassi et al. (2021) have eliminated most of the discrepancy. This is a major development to forest-based climate actions and carbon accounting (Ogle and Kurz 2021). The simple message from it is that at least for now, we can determine whether the forest biomass and carbon dynamics are driven by human actions and whether we can conduct forest carbon accounting based directly on the NFI information and the delineation provided therein. So long as some sort of management is present, the FES and thus its biomass and soil carbon pools can be included in the accounting; otherwise, if the FES is unmanaged, its biomass and soil carbon pools should be excluded.

## *Reference levels*

It should be emphasized that REDD+ and non-REDD+ situations feature different baselines or reference levels. For non-REDD+ countries that have internal targets of emission reduction and removal in their NDCs, the reference levels for these actions are set at a particular point of time, expressed as tons of $CO_2$ equivalent a year. Using the China example again, given its choice of 2005 as the base year for its NDC, the amounts of GHG emissions and those of carbon sequestration in the FESs and storage in HWPs in 2005 are thus the reference levels. Emission reduction or removal by a future year (say, 2030) are determined against these reference levels, as shown in Box 1.3 later.

For REDD+ countries, on the other hand, some kind of counterfactual or business as usual (BAU) scenario for emission reductions from deforestation and forest degradation (i.e., REDD), or removals from afforestation and reforestation, improved forest management, and enhanced HWP carbon stock (i.e., +), must be established first before the offsetting credits and payments can be assessed (Roopsind et al. 2019, Nepstad et al. 2013)[4]. Here, reference levels are for a designated period, against which the emissions and removals from a results period will be compared. The challenge is to identify the trend lines for the REDD+ activities accurately, especially given the lack of reliable historical measurement and monitoring and/or the variation of existing observations of the forest condition over time. Considering the NDC architecture of the Paris Agreement and the feasibility of developing these reference levels, thus, the UNFCCC is primarily interested in national forest reference levels for a REDD+ country, such that these levels will maintain consistency with the country's greenhouse gas inventory estimates, as detailed in the UNFCCC REDD+ Web Platform information.

Of course, as an interim measure, subnational forest reference levels, can be developed and used, and the COP recognizes the importance and necessity of adequate and predictable financial and technology support for developing such reference levels. Once developed and submitted, they will then be assessed by a technical review process organized by the COP. It has been reported that identifying the reference levels for REDD+ national and subnational jurisdictions, let alone for the local projects, can be difficult, and inaccuracy and even leakage may occur (Mertz et al. 2018, West et al. 2020). I will deliberate this and other problems further in the next chapter when expounding the different facets of governing forest sector climate solutions, with the empirical evidence and discussion to be presented in Chapters 7 and 9.

Another related complication is the fact that some countries, such as Indonesia and Brazil, have made both unconditional and conditional commitments—the former type to be fulfilled without results-based payments from international sources while the latter type to be fulfilled only with results-based payments from international sources. In theory, a common set of references levels for emission and removal should be developed and used, especially because the commitments are often made on a relative basis. In practice, however, the reference levels for the conditional commitments may be subject to more scrutiny and dispute than those for the unconditional levels. The COP will need to work out a solution to this issue in consultation and negotiation with the relevant Parties, as well as its technical assessment apparatus.

*Carbon accounting*

A mitigation action can only be accounted for when it represents a reduction of anthropogenic emissions relative to an established *baseline*. As alluded, establishing a benchmark is non-trivial to forest sector actions. It is thus a key first step to define and agree upon a Party's benchmark, against which the *additionality* of those actions is determined. In a net–net accounting, for example, total net emissions/removals in the base year are subtracted from total net emissions/removals in the accounting period to find the total amount of credits/debits resulting from these actions (UNFCCC 2020). Obviously, this method is well-suited to non-REDD+ Parties (e.g., China, India, and the US) but not necessarily REDD+ countries (e.g., Brazil, Indonesia, and the Republic of Congo).

In addition to credit taking, a results-based payment scheme from REDD+ actions may be pursued by a developing country through international cooperation of carbon offsetting with a developed country and/or other organizations, in which case the UNFCCC is focused on mitigation beyond BAU (FAO 2016). That implies that the only actions that should count are those that reduce atmospheric $CO_2$ above and beyond what would occur in the absence of the executed action(s), which is another way of determining *additionality* (Sedjo and Sohngen 2012). Given this definition, reduced emissions (or increased removals) by forest sector actions are not always additional (Mertz et al. 2018, Roopsind et al. 2019, Coffield et al. 2022).

Since carbon sequestered in trees is eventually emitted back to the atmosphere after they are harvested, or lost to natural mortality or disturbances, any forest-based mitigation action can lead to non-permanence if not well designed. *Permanence* necessitates that carbon sequestered/stored that has entered an accounting system will not be released over a sufficiently long period (FAO 2016). As a result, forest sector actions inevitably entail a long-term land-use or management commitment, which can make some of them less attractive or less feasible (e.g., afforestation and reforestation projects where land may be converted back to non-forest uses).

In general, the duration of an action should be commensurate with the life of an NDC; certainly, it must be longer and more consistent than what was practiced under the Kyoto Protocol notions of temporary or long-term certified emissions reduction (FAO 2016). In fact, the IPCC (2022) and other organizations have indicated that a long-term climate change assessment ought to cover a time horizon at least until the end of this century (Ou et al. 2021). But this condition is no longer a major hurdle once we look at forest-based carbon sequestration/storage from the jurisdictional perspective, as noted earlier.

Also, in addition to changes in land use practices for sequestering carbon, the possibility of trade between different jurisdictions (domestic or international) means that we must account for the carbon flow associated with the product flow. If one jurisdiction imports more wood products than it exports, it can lead to the problem of carbon leakage as well. *Leakage* refers to the potential outcome that land-use changes and/or wood product flows across jurisdictions may result in carbon release elsewhere, i.e., a displacement of the emissions (Pan et al. 2020, Sedjo and Sohngen 2012). To fulfill its NDC, to take carbon credits, or to access results-based payments, a Party must also put in place a national strategy, a national (and sub-national) forest monitoring system, and a report on how safeguards are addressed (FAO 2016). In case certain practices are inconsistent with these requirements, they must be redressed before national progress tracking or global stock taking gets underway. Box 1.3 summarizes the framework of forest carbon accounting for China and the estimated outcomes during 2006–2020.

### Box 1.3 The accounting framework and outcome for China

Given China's NDC, we should take its emission, sequestration, and storage in 2005 as the reference levels; additionality is then measured as the reduced emission or increased removal against these reference levels. Since the 2020 carbon emission has not yet been reported, we have estimated that the total carbon emission during 2006–2020 would reach 36.792–37.031 petagrams (Pg). Meanwhile, the forest area has gained by $47.6 \times 10^6$ ha, and the average unit stock volume increased to 80.59 m$^3$/ha during 2006–2020. In combination, these changes result in a total forest biomass increase of $5.07 \times 10^9$ m$^3$. Carbon sequestered in forest biomass would amounts to 2.592 Pg, offsetting 7.00% of the emissions, which is very close to what was predicted by Yin et al. (2010). Adding carbon storage in HWPs (0.491 Pg), we get an offsetting ratio of 8.33%. With a 30% FES carbon in soil, the offsetting ratio becomes11.77% (Hou and Yin 2022).

## Closing remarks

The purpose of this chapter was to deliberate the processes, principles, and rules relevant to forest sector climate actions considering the Paris Agreement's NDC structure. Along the way, I also shared my perspectives of and solutions to some of the tough problems likely to be encountered in implementing certain forest sector actions. I think that this kind of exploration is broadly interesting and informative, and I hope that my efforts will be well received by readers. Certainly, I feel that this opening chapter has laid a clear, concrete foundation on which I will be able to shape and form the other chapters of this book.

In summary, corresponding to the actions and targets specified in the NDC, a Party must select appropriate indicators and use proper assumptions and methodologies to track progress in meeting its targets. Then, it becomes feasible to conduct carbon accounting, as well as stock taking, following the relevant principles and rules, such as additionality, permanence, and leakage measured against an established baseline or reference level. However, there are distinct reference levels under different circumstances. The Party must possess consecutive GHG inventories for emissions and removals to determine, communicate, and update its NDC.

There is limited concern over the availability of GHG emission inventories, while there are large gaps between the available information and what is needed for accounting forest sector actions. That is, Parties tend to have reliable data for forest stock volume, from which the amount of carbon sequestered in the biomass can be estimated; however, they may not have good information on the carbon fluxes associated with the forest soil pool, urban trees and forests, or trees outside of forests elsewhere (FAO 2020b). Improved measurement and monitoring are needed if a Party desires to include these carbon pools as part of the offsetting credits in its NDC accounting.

At the same time, it will be difficult to get the results-based payments scheme carried out at scale, because, on the one hand, it takes time and effort to resolve the challenges of determining the reference levels of REDD+ projects and programs and thus their carbon additionality, as well as leakage and permanence. On the other hand, it has been a struggle to obtain the promised international finance. The 2009 pledge by developed nations to mobilize 100 billion USD a year by 2020 in climate financing to developing nations has not been delivered (Roberts et al. 2021). A large portion of those funds (~10–15 billion USD a year) is supposed to be allocated for the REDD+ initiative (Groom et al. 2022). While the G7 countries reiterated their intention to keep the promise at their meeting in the UK, it is unclear if and how they will fulfill that promise (NCC Editorial 2021). Space limit does not permit me to get into carbon finance as well as pricing, which I will address in Chapter 7.

Furthermore, if forest and other terrestrial ecosystems will play a truly significant role in reducing and ultimately neutralizing carbon emissions (Roe et al. 2019, Griscom et al. 2017), then it is vital to protect FESs from degradation and destruction, especially those primary natural forests in the tropics. Meanwhile, Parties must do more to improve their forest conditions as reflected in the structure, quality, and growth by adopting adequate silvicultural practices. In Box 1.4, I show the current condition of China's FESs and argue for undertaking extensive stand treatments to boost forest growth and domestic timber supply. As stand structure improves and growth accelerates, the country's FES productivity and functionality will be enhanced, leading to greater amounts of carbon to be sequestered and other services generated (Liu and Yin 2012). More consumption of domestic wood can in turn reduce China's ecological footprint overseas and increase carbon stored in HWPs. Together, these shifts will go a long way toward meeting the goals of national mitigation and maintaining biodiversity and the provision of FES co-benefits (Mallapaty 2020, Hou et al. 2019).

> **Box 1.4 China's current forest conditions**
>
> In 2018, the mean stocking level of arbor forests was 95m$^3$/ha and that of planted forests was 59.30m$^3$/ha, the mean annual increment of all forests was 4.73m$^3$, and the proportion of young and middle-aged forests accounted for over 64% of all forests (Ke et al. 2020). These figures are far below or deviate from international counterparts, as well as China's forestland production potentials. With the limited opportunity for further area expansion, it will have to rely on improved management practices for enhancing forest growth and carbon offsetting (Li et al. 2016).

Finally, my analysis in this chapter calls for reorienting the inquiry into forest carbon policy and economics from a predominantly disaggregated, project-based paradigm to a more aggregated, program-oriented one, such that it will be more congruent with and conducive to the NDC architecture and a jurisdictional approach. Of course, the former kind of work remains useful and can effectively complement the latter type. Because it may not allow the forest growth process to be continuous, however, the former tends to treat timber production and carbon sequestration as less compatible processes and thus make it hard to abide by the PA requirements (Li et al. 2020).

Also, it is not necessarily a choice of either timber or carbon in reality; rather, it is a matter of both in many cases. Likewise, timber and carbon may not be substitutes since their costs and prices can diverge over time and space. It will be interesting to assess the tradeoffs between these (and other) ecosystem services and explore their efficient, effective, and equitable provision. I will take up some of these considerations in Chapter 4.

## Notes

1 See a definition of REDD+ in the Introduction.
2 In my opinion, the relevance of considering carbon stored in HWPs and its linkage with carbon sequestered in FESs makes it more appropriate and meaningful to refer to these actions together as forest sector climate actions (or solutions), instead of forest-based actions as commonly used in the literature.
3 According to Liu and Yin (2012) and Guo et al. (2014), arbor stands, economic forests, and bamboo groves are treated as regular types of forests in China, whereas the other elements—woodlands, shrub-lands, and trees on non-forest land including four-side greening and scattered trees—are regarded as trees outside of forests, or TOFs.
4 In principle, results-based payment initiatives of the public domain can be carried out in the forest sector of a non-REDD+ country as well. But the few examples I am aware of are the emission trading schemes (ETS) of California, China, and New Zealand, though all have their challenges in terms of either the "results" or "payment," or even both. For instance, California's ETS allows no more than 8% of the capped emissions to be offset by various means of sequestration and storage, with payment at the ongoing trade price of allowances (Coffield et al. 2022). I will come back to this issue in Chapters 7 and 9.

## References

Boyd, W., Stickler, C., Duchelle, A.E., Seymour, F., Nepstad, D. et al. 2018. *Jurisdictional Approaches to REDD+ and Low Emissions Development: Progress and Prospects.* Washington, DC: World Resources Institute.

Chagas, T., Galt, H., Lee, D., Neeff, T., Streck, C. 2020. *A Close Look at the Quality of Redd+ Carbon Credits.* Available at: www.climatefocus.com.

Coffield, S.R., Vo, C.D., Wang, J.A., Badgley, G. et al. 2022. Using remote sensing to quantify the additional climate benefits of California forest carbon offset projects. *Global Change Biology* 28: 6789–6806. https://doi.org/10.1111/gcb.16380.

Energy Foundation China. 2020. *Synthesis Report 2020 on China's Carbon Neutrality: China's New Growth Pathway: from the 14th Five Year Plan to Carbon Neutrality.* Beijing, China. Available at: www.efchina.org/Reports-en/report-lceg-20201210-en.

Fang, J.Y., Chen, A., Peng, C., Zhao, S., Ci, L. 2001. Changes in forest biomass carbon storage in China between 1949 and 1998. *Science* 292: 2320–2322.

Food and Agriculture Organization of the United Nations (FAO). 2016. *Forestry for a Low-Carbon Future: Integrating Forests and Wood Products in Climate Change Strategies* (FAO Forestry Paper 177). Rome, Italy.

Food and Agriculture Organization of the United Nations (FAO). 2020a. *Trees Outside Forests: Towards a Better Awareness.* Rome, Italy.

Food and Agriculture Organization of the United Nations (FAO). 2020b. *Global Forest Resource Assessment Report.* Rome, Italy.

Fischer, R., Hargita, Y., Günter, S. 2016. Insights from the ground level? A content analysis review of multi-national REDD+ studies since 2010. *Forest Policy and Economics* 66: 47–58.

Galik, C., Murray, B.C., Mitchell, S., Cottle, P. 2016. Alternative approaches for addressing non-permanence in carbon projects: an application to afforestation and reforestation under the Clean Development Mechanism. *Mitigation and Adaptation Strategies for Global Change* 21: 101–118.

Geng, A., Chen, J., Yang, H.Q. 2019. Assessing the greenhouse gas mitigation potential of harvested wood products substitution in China. *Environmental Science & Technology* 53 (3): 1732–1740.

Grassi, G., House, J., Kurz, W.A. et al. 2018. Reconciling global-model estimates and country reporting of anthropogenic forest CO2 sinks. *Nature Climate Change* 8: 914–920.

Grassi, G., Stehfest, E., Rogelj, J., van Vuuren, D., Cescatti, A. et al. 2021. Critical adjustment of land mitigation pathways for assessing countries' climate progress. *Nature Climate Change* 11: 425–434.

Gren, I-M., Zeleke, A.A. 2016. Policy design for forest carbon sequestration: A review of the literature. *Forest Policy and Economics* 70: 128–136.

Griscom, B.W., Adams, J., Ellis, P.W., Houghton, R.A., Lomax, G., Miteva, D.A., Schlesinger, W.H., Shoch, D., Siikamäki, J.V., Smith, P. et al. 2017. Natural climate solutions. *PNAS* 114 (44): 11645–11650.

Groom, B., Palmer, C., Sileci, L. 2022. Carbon emissions reductions from Indonesia's moratorium on forest concessions are cost-effective yet contribute little to Paris pledges. *PNAS* 119 (5): e2102613119.

Guo, Z.D., Hu, H.F., Pan, Y.D., Birdsey, R.A., Fang, J.Y. 2014. Increasing biomass carbon stocks in trees outside forests in China over the last three decades. *Biogeosciences* 11: 4115–4122.

Hou, J.Y., Yin, R.S., Wu, W.G. 2019. Intensifying forest management in China: What does it mean, why, and how? *Forest Policy and Economics* 98: 82–89.

Hou, J.Y., Yin, R.S. 2022. How significant a role can China's forest sector play in decarbonizing its economy? *Climate Policy* 23(2): 226–237. doi:10.1080/14693062.2022.2098229.

International Panel on Climate Change (IPCC). 2006. *Guidelines for National Greenhouse Gas Inventories.* Available at: www.ipcc.ch/report/2006-ipcc-guidelines-for-national-greenhouse-gas-inventories/.

International Panel on Climate Change (IPCC). 2019. *2019 Refinement to the 2006 IPCC Guidelines for National Greenhouse Gas Inventories.* Available at: www.ipcc.ch/report/2019-refinement-to-the-2006-ipcc-guidelines-for-nationalgreenhouse-gas-inventories/.

International Panel on Climate Change (IPCC). 2020. *Climate Change and Land: Summary for Policymakers.* Available at: www.ipcc.ch/srccl/.

International Panel on Climate Change (IPCC). 2022. *Climate Change 2022: Impacts, Adaptation and Vulnerability.* Working Group II Contribution to the Sixth Assessment Report.

Johnston, C.M.T., Radeloff, V.C. 2019. Global mitigation potential of carbon stored in harvested wood products. *PNAS* 116: 14526–14531.

Juutinen, A., Ahtikoski, A., Lehtonen, M., Mäkipää, R., Ollikainen, M. 2018. The impact of a short-term carbon payment scheme on forest management. *Forest Policy and Economics* 90: 115–127.

Ke, S.F., Qiao, D., Yuan, W.T., He, Y.J. 2020. Broadening the scope of forest transition inquiry: What does China's experience suggest? *Forest Policy and Economics* 118 (102240).

Li, P., Zhu, J., Hu, H., Guo, Z., Pan, Y., Birdsey, R., Fang, J.Y. 2016. The relative contributions of forest growth and areal expansion to forest biomass carbon. *Biogeosciences* 13: 375–388.

Li, X.Y., Lu, G., Yin, R.S. 2020. Research trends: Adding a profit function to forest economics. *Forest Policy and Economics* 113 (102133).

Liu, P., Yin, R.S. 2012. Sequestering carbon in China's forest ecosystems: Potential and challenges. *Forests* 3: 417–430.

Luyssaert, S., Ciais, P., Piao, S.L. et al., in press. 2010. The European carbon balance. Part 3. Forests. *Global Change Biology* 16: 1429–1450. doi:10.1111/j.1365-2486.2009.02056.x.

Mallapaty, S. 2020. How China could be carbon neutral by mid-century? *Nature* 586: 482–483.

Mertz, O., K. Grogan, D. Pflugmacher et al. 2018. Uncertainty in establishing forest reference levels and predicting future forest-based carbon stocks for REDD+. *Journal of Land Use Science* 13: 1–15.

Nature Climate Change (NCC) Editorial. 2021. Enhancing commitments. *Nature Climate Change* 11: 457.

Nepstad, D.C., Boyd, W., Stickler, C.M., Bezerra, T., Azevedo, A.A. 2013. Responding to climate change and the global land crisis: REDD+, market transformation and low-emissions rural development. *Philosophical Transactions of the Royal Society B* 368 (1619): 20120167.

National Development and Reform Commission (NDRC). 2015. *Enhanced Actions on Climate Change: China's Intended Nationally Determined Contributions* (June 30). Beijing, China.

National Forestry and Grassland Administration of China (NFGA). 2019. *A Summary of the China's Forest Condition Derived from the Ninth National Inventory*. Beijing, China.

Nordhaus, W. 2021. *The Spirit of Green*. New Jersey: Princeton University Press.

Ogle, S.M., Kurz, W.A. 2021. Land-based emissions. *Nature Climate Change* 11https://doi.org/10.1038/s41558-021-01040-7.

Ou, Y., Iyer, G., Clarke, L., Edmonds, J. et al. 2021. Can updated climate pledges limit warming well below 2°C? *Science* 6568: 693–695.

Pan, W.Q., Kim, M.K., Ning, Z., Yang, H.Q. 2020. Carbon leakage in energy/forest sectors and climate policy implications using meta-analysis. *Forest Policy and Economics* 115 (102161).

Phelps, J., Webb, E.L., Agrawal, A. 2010. Does REDD+ threaten to recentralize forest governance? *Science* 328: 312–313.

Piao, S.L., Fang, J.Y., Ciais, P., Peylin, P., Huang, Y., Sitch, S., Wang, T. et al. 2009. The carbon balance of terrestrial ecosystems in China. *Nature* 458: 1009–1013.

Republic of Indonesia. 2022. *Enhanced Nationally Determined Contribution Submitted to UNFCCC Secretariat*. Available at: https://unfccc.int/reports.

Roberts, J.T., Weikmans, R., Robinson, S., Ciplet, D., Khan, M., Falzon, D. 2021. Rebooting a failed promise of climate finance. *Nature Climate Change* 11: 180–182.

Roe, S., Streck, C., Obersteiner, M., Frank, S., Griscom, B., Drouet, L. et al. 2019. Contribution of the land sector to a 1.5°C world. *Nature Climate Change* 9: 817–828. doi:10.1038/s41558-019-0591-9.

Roopsind, A., Sohngen, B., Brandt, J. 2019. Evidence that a national REDD+ program reduces tree cover loss and carbon emissions in a high forest cover, low deforestation country. *PNAS* 116 (49): 24492–24499.

Sedjo, R., Sohngen, B. 2012. Carbon sequestration in forests and soils. *Annual Review of Resource Economics* 4:127–144.

Seymour, F., Busch, J. 2016. *Why Forests? Why Now? The Science, Economics, and Politics of Tropical Forests and Climate Change.* Washington, DC: Brookings Institution Press.

Songwe, V., Stern, N., Bhattacharya, A. 2022. *Finance for Climate Action: Scaling Up Investment for Climate and Development.* London: Grantham Research Institute on Climate Change and the Environment, London School of Economics and Political Science.

Thamo, T., Pannell, D.J. 2016. Challenges in developing effective policy for soil carbon sequestration: perspectives on additionality, leakage, and permanence. *Climate Policy* 16: 973–992.

United Nations Framework Convention on Climate Change (UNFCCC) Secretariat. 2015. *Synthesis Report on the Aggregate Effect of the Intended Nationally Determined Contributions.* Available at: https://unfccc.int/resource/docs/2015/cop21/eng/07.pdf.

United Nations Framework Convention on Climate Change (UNFCCC) Secretariat. 2020. *Reference Manual for the Enhanced Transparency Framework under the Paris Agreement.* Bonn, Germany.

United Nations Framework Convention on Climate Change (UNFCCC). *REDD+ Web Platform.* Available at: https://redd.unfccc.int/fact-sheets/forest-reference-emission-levels.html.

Van der Gaast, W., Sikkema, R., Vohrer, M. 2018. The contribution of forest carbon credit projects to addressing the climate change challenge. *Climate Policy* 18 (1): 42–48.

Wang, Y.F., Li, L.Y., Yin, R.S. 2021. A primer on forest carbon policy and economics under the Paris Agreement. *Forest Policy and Economics* 132 (102595).

West T.A.P., Borner J., Sills E.O., Kontoleon A. 2020. Overstated carbon emission reductions from voluntary REDD+ projects in the Brazilian Amazon. *PNAS* 117 (39): 24188–24194.

Wunder, S., Duchelle, A.E., Sassi, C., Sills, E.O., Simonet, G., Sunderlin, W.D. 2020. REDD+ in theory and practice: How lessons from local projects can inform jurisdictional approaches. *Front. For. Glob. Change* 3: 11. doi:10.3389/ffgc.2020.00011.

Xinhua. 2020. Remarks by Chinese President Xi Jinping at the Climate Ambition Summit. Available at www.xinhuanet.com/english/2020-12/12/c_139584803.htm.

Yin, R.S., Sedjo, R.A., Liu, P. 2010. The potential and challenges of sequestering carbon and generating other services in China's forest ecosystems. *Environmental Science & Technology* 44: 5687–5688.

Zeng, W.S., Tomppo, E., Healey, S.P., Gadow, K.V. 2015. The National Forest Inventory in China: History, results, and international context. *Forest Ecosystems* 2 (1): https://doi.org/10.1186/s40663-015-0047-2.

# 2 Understanding and undertaking jurisdictional approaches to governing forest climate actions

**Introduction**

Unlike the explicit project orientation of the Kyoto Protocol, the nationally determined contribution (NDC) architecture of the Paris Agreement underlines a jurisdictional approach to international climate governance—making and executing pledges by administrative bodies of a national government, with the obligations of subordinate bodies being nested within the national pledge (von Essen and Lambin 2021, DeFries et al. 2022). Meanwhile, the UNFCCC tracks the progress that has been made by a national government and assesses to what extent the pledged commitments will collectively meet the global temperature targets (UNFCCC 2020).

Of course, as a transitional step, subnational jurisdictions can independently commit themselves to certain international responsibilities if their positions on climate change mitigation deviate from those of the national government. A case in point is the Governor's Climate and Forests Task Force—a transnational network of subnational authorities—created to further the development of jurisdictional REDD+ programs (Parrotta et al. 2022).[1] However, this short-term, interim measure should not be confused with the long-term, fundamental mechanism of NDCs (FAO 2016). The aim of this chapter is to examine the distinctions, strengths, and executions of alternative approaches to forest climate governance from a comparative perspective.

Compared to project-based approaches, jurisdictional ones have certain advantages. As briefly noted in Chapter 1, jurisdictional approaches provide a greater likelihood for countries to carry out necessary actions and assume their responsibilities. Also, they make it more practical for the carbon accounting principles to be materialized, and jurisdictions tend to have a stronger capability of measurement and MRV (i.e., monitoring, reporting, and verification) in carrying out their commitments. Adopting jurisdictional approaches could thus have made what we learned under the Kyoto Protocol or from the experience of voluntary markets less adequate. Nonetheless, our current knowledge of forest climate governance, including carbon accounting, offsetting, and crediting, has been somehow limited within the domain of carbon projects (e.g., Nepstad et al. 2013, Galik et al. 2016) or distorted by

DOI: 10.4324/9781003436652-3

what has been unfolding on voluntary markets (e.g., DeFries et al. 2022, Bossio et al. 2020).

To further this point, 90 percent of the "second generation" NDCs—those new or updated ones that were submitted to the UNFCCC by the end of 2020—contain Land Use, Land-Use Change and Forestry (LULUCF) actions. (Refer to the UN REDD+ Web Platform at https://redd.unfccc.int/ for more details.) In comparison, the UN REDD+ program itself now covers just 65 tropical countries. Obviously, forest sector climate actions have gone far beyond the scope and significance of the REDD+ initiative. Likewise, while active and rapidly evolving, voluntary markets, including transactions in the LULUCF arena, represent a small portion of the existing, let alone the expected, size of world carbon markets (Ecosystem Marketplace 2021).

In fact, other than a few large bilateral and multilateral assistance agreements, it has been difficult to get the results-based payments scheme for REDD+ projects carried out at scale (Parrotta et al. 2022). This is partly because it takes time and effort to resolve the challenges of determining their reference levels and thus carbon additionality, leakage, and permanence. Additionality concerns whether and to what extent emission reduction and removal occur due to certain intervention(s); permanence refers to whether the benefits of emission reduction and removal may be reversed; and leakage refers to whether reductions in one place are displaced by emissions to another place (Thamo and Pannell 2016, Wang et al. 2021).

At the same time, it has been a struggle to obtain the promised international finance. The 2009 pledge by developed nations to mobilize US$100 billion a year by 2020 in climate financing to developing nations has not been delivered (Roberts et al. 2021), and a large portion of those funds (~ US$10–15 billion a year) is supposed to be allocated for the REDD+ scheme (Groom et al. 2022). On the other hand, there was an early explosion of local, non-jurisdictional REDD+ projects in response to the UNFCCC 2007 call for "demonstration activities," as well as perceived funding opportunities. As Wunder et al. (2020) put mildly, implementing conditional payments ran into both a supply- and demand-side problem, which seems to jointly explain the discrepancy between REDD+ theory and practice.

In short, certain processes, principles, and practices pertinent to forest climate governance have not been well understood, let alone properly implemented. For instance, while most scholars and policymakers are aware of the principles of additionality, permanence, and leakage avoidance for the accounting and evaluation of forest sector climate solutions, the question remains whether these principles should be followed primarily at the national (or subnational) level or at the individual project level.

From the global perspective, it seems more appropriate and relevant to examine how these principles are followed by national jurisdictions in line with the NDC architecture of the Paris Agreement. Moreover, it would be unimaginable for the international community to check how these principles are followed at the individual project level, given the likely number of projects to be carried out within each jurisdiction and the heterogeneity of local circumstances (West et al. 2020). Certainly, this does not mean that a national jurisdiction should not consider these principles when it designs, executes, or evaluates forest carbon programs and projects within its borders. This is actually the subsidiarity principle—social and political issues should be dealt with at the most immediate level that is consistent with their resolution—that Professor Nordhaus (2021) has stressed.

Therefore, it has become crucial to harmonize our past experiences and lessons with the changed landscape of climate governance regarding emission reduction and removal (IPCC 2022), and it is essential to resolve the challenges facing the international community in executing forest sector nature-based solutions (NbSs) to the climate crisis (Oldfield et al. 2022). Doing so at this time is of particular importance to advancing the global climate agenda (Roe et al. 2019). Therefore, following a careful literature review, I will deliberate the differences between project- and jurisdiction-based approaches and the advantages of the latter ones to forest sector climate governance, before discussing the policy and practice implications of my review and assessment in this chapter.

## Literature review

Scholars have deliberated and debated the various facets of governing forest sector climate actions. Murray et al. (2006) examined the challenges of forest carbon accounting with respect to the principles of additionality, permanence, and no leakage in the context of soil carbon sequestration and storage from a project orientation. Similarly, Galik et al. (2016) explored the alternatives for addressing non-permanence in afforestation and reforestation projects under the Clean Development Mechanism. Given the Kyoto Protocol's project orientation, however, these discussions appear to be dated and hence inadequate in a post-Kyoto Protocol world (Parrotta et al. 2022). Recently, there have been more substantive explorations on how to govern forest-based mitigation actions in the LULUCF arena (e.g., DeFries et al. 2022, Parrotta et al. 2022, Bossio et al. 2020), and there has even been a convergence toward jurisdictional approaches in the deliberation (von Essen and Lambin 2021, Ecosystem Marketplace 2021).

For instance, Parrotta et al. (2022) provided a timely examination of the role of REDD+ in global forest governance and its translation into practice at various scales. In fact, REDD+ was conceived as a voluntary, 'nationally driven' mechanism for high-income countries (as listed in Annex 1 of the UNFCCC) to pay low- and middle-income countries (non-Annex 1 Parties) for reducing forest sector emissions. This mechanism is thus rested on strong international cooperation. In Chapter 2 of Parrotta et al. (2022), McDermott et al. applied a 'political ecology' lens to highlight how the dynamics of power that shape REDD+ playout across the principal sources of authority. As REDD+ has evolved, however, so too have some global negotiations, initiatives, and programs aimed at addressing climate change, deforestation, land degradation, and other challenges. For example, many countries with no REDD+ activity have been promoting their forest sector NbSs. In the words of McDermott et al. (2022), these developments have not just blurred the boundaries of REDD+ and other forest sector NbSs but have also carried major implications for the governance, including funding, the scope of activities, and on-the-ground interventions.

Consequently, REDD+ is no longer inclusive of the full efforts of the forest sector emission reduction and removal. In fact, REDD+ projects themselves have been shifting toward jurisdictional regimes (Ecosystem Marketplace 2021, UNFCCC 2022), which are viewed to be crucial to the successful execution of NbSs at the national level (Irawana et al. 2019). Further, while the 'political ecology' perspective is constructive and convenient, it is essential to move beyond the power dynamics of stakeholders and make the governance system operational. To that end, it is worthwhile to examine issues related to the delivery of the mitigation actions promised in the NDCs, the compatibility with the Paris Agreement accounting principles, the capability of measurement and MRV, and the alleviation of transaction risk and cost (Chagas et al. 2020, Wang et al. 2021).

Noting that roughly half of all credits issued from 2000 to 2021 on the voluntary carbon markets were related to land use, mostly from forest projects, DeFries et al. (2022) asserted that demand for credits on these markets would be poised to surge as corporations like Microsoft and Amazon implement net-zero commitments. This is also the view held by Bossio et al. (2020) when articulating the role of soil carbon in NbSs. In addition, the attention voluntary carbon markets have garnered has to do with the fervor to generate carbon credits in these markets, coupled with the slow expansion of compliance markets (Parrotta et al. 2022, Wang et al. 2021). Nonetheless, the overall size of the carbon market remains small. Only ~25% of global GHG emissions are subject to any price at all, and 75% of emissions that are subject to a price are charged at less than $10 per ton of carbon dioxide equivalent (World Bank 2021). Moreover, REDD+ and market-based forest project finances have accounted for only ~3% of the total expenditure spent on climate change mitigation and adaptation (Midkiff et al. 2022). Tackling the governance issues thus requires an understanding of the acute need for more ambitious efforts of emission reduction and removal, which in turn calls for a more inclusive perspective and a better integrated framework for both accounting and assessment.

It is also worth noting that the journal of *Frontiers in Forests and Global Change* published a Special Section of nine articles in 2020 titled Jurisdictional Approaches to Sustainability in the Tropics. Included are "REDD+ in Theory and Practice: How Lessons From Local Projects Can Inform Jurisdictional Approaches?" by Wunder et al. (2020), "Lessons for Jurisdictional Approaches From Municipal-Level Initiatives to Halt Deforestation in the Brazilian Amazon" by Brandão et al. (2020), and "The Jurisdictional Approach in Indonesia: Incentives, Actions, and Facilitating Connections" by

Seymour et al. (2020). These articles have contributed to improved knowledge of what we can learn from the early experience of REDD+ projects and how we may scale them up to jurisdictional programs. As informative and impressive as these and other studies may be, however, none of them ever went beyond the domain of REDD+.

Similarly, the definition of additionality by DeFries et al. (2022)—whether interventions to reduce emissions or store carbon would occur in the absence of revenue from the (voluntary) market—is germane to REDD+ and other results-based payments projects. However, it is not necessarily applicable to an array of forest sector actions undertaken by countries without REDD+, like the US, China, and India. This is because forest sector NbSs in REDD+ and non-REDD+ countries feature different baselines as well as rules of carbon accounting. I will elaborate this point in the next section; here, it suffices to state that it is vital to put the relevant issues in a proper context to build a more coherent understanding of them.

## Distinctions between alternative approaches

There are multiple distinctions between project-based and jurisdictional approaches to governing forest sector climate actions. First, different approaches may not have the same baselines or financing mechanisms. For countries engaged in REDD+ and other similar projects of results-based payments, the baseline GHG emission reductions from deforestation and forest degradation (i.e., REDD), and removals through afforestation and reforestation, improved forest management, and enhanced carbon stock (i.e., +) are the outcomes of the so-called business as usual (BAU) scenarios that must be established before the offsetting credits and thus payments can be assessed (Roopsind et al. 2019, Nepstad et al. 2013). Here, the reference levels are for a reference period, against which the emissions and removals from a results period will be compared, as detailed on the UN REDD+ Web Platform.

The challenge is to identify the trend lines for the REDD+ activities accurately, especially given the lack of reliable historical measurement and monitoring and/or the variation of existing observations of forest condition over time. Any emissions (removals) below (above) the corresponding baseline scenarios are deemed additional and should thus be recognized and rewarded with payments (McDermott et al. 2022). In contrast, for non-REDD+ countries that undertake forest sector NbSs, while their primary responsibilities are emission reduction, they are allowed to pursue emission removal as well. These actions are all counted in reference to the same base year—total emission reduction and removal in the base year are subtracted from their counterparts in the future accounting period to find the total additional number of credits or debits resulting from these actions (UNFCCC 2020). So, the baseline or reference level is different from the counterfactual determined for a REDD+ project or program (Wang et al. 2021).

A further complication is the fact that some countries, such as Indonesia and Democratic Republic of Congo, have made both unconditional and conditional commitments—the former type will be fulfilled without results-based payments from international sources while the latter type will be fulfilled only with results-based payments from international sources. In theory, a common set of reference levels for emission and removal should be developed and applied, especially given the commitments are often made in a similar way—say, an unconditional emission reduction of 29% by 2030 against the 2005 baseline or 41% if conditional emission reduction is included as well, as pronounced by Indonesia in its original NDC (Republic of Indonesia 2022). In practice, however, the reference levels for the conditional commitments may be subject to more scrutiny and dispute than those for the unconditional pledges.

Moreover, while carbon sequestered or stored by forest sector actions under the NDCs can be traded internationally, that is not a prerequisite for non-REDD+ countries. They carry these actions out mostly domestically, with revenues generated from a carbon pricing mechanism or a combination of public and private finance (Midkiff et al. 2022). As such, we can expect that these actions are more likely to be executed as part of a jurisdictional program rather than in individual projects. Moreover, the pace of their adoption may well surpass that of REDD+ projects, given the distinct operating and funding mechanisms.

As a matter of fact, von Essen and Lambin (2021) reported that over the past decade, jurisdictional approaches to sustainable resource use have attracted increasing attention as a potential alternative to traditional conservation strategies. Often, these approaches operate within formal administrative boundaries and seek to establish policies and practices that apply to all stakeholders. The authors also highlighted that these jurisdictions covered ~40% of global tropical forests, with most undergoing higher-than-average deforestation rates. These are indications consistent with the overall characterization of the REDD+ program, which "has a national rather than project scope, with a subnational approach possible as an interim measure to lead towards the national scale, in a way that non-permanence and leakage risks can be consistently and more cost-effectively addressed" (FAO 2016).

Furthermore, project-based approaches have salient limitations. Under the Kyoto Protocol's Clean Development Mechanism, an afforestation or reforestation project must demonstrate that additional effort in terms of financial investment or other resources is necessary for it to be implemented (Fischer et al. 2016). Participants can select one of the two types of crediting periods and their lengths:

i    a single fixed crediting period of up to thirty years;
ii    a renewable crediting period of up to twenty years which can be renewed twice.

The maximum length of the crediting period of an afforestation/reforestation project activity, if successfully renewed both times, can be sixty years (FAO 2016). But it is hard for such a project to be permanent.

In addition, the scale of afforestation and reforestation projects under the Clean Development Mechanism is too small to make a difference (UNFCCC 2013). DeFries et al. (2022) already critiqued that a shortcoming of the current project-based carbon market is its small-scale, piecemeal approach through individual projects. As a result, it is difficult to foster transitions to low-carbon land management over larger areas for a meaningful aggregate impact. As jurisdictional approaches become more common, however, it is possible for them to overcome the shortcomings of a project-based approach.

Mehling et al. (2018) further argued that one promising avenue to incentivize countries to raise climate mitigation ambitions over time is not just to have jurisdictional approaches to climate governance front and center but also to link different policies such that emission reduction and removal in one jurisdiction can be counted toward commitments of another jurisdiction. Linkage is key because it can reduce the costs of achieving a given emission reduction and removal objective. Lower costs, in turn, may contribute politically to embracing more ambitious objectives. In a world where marginal abatement cost varies widely, linkage improves cost-effectiveness by allowing one jurisdiction to finance emission reduction and removal in another with relatively lower costs while allowing the former to count the emission reduction and removal toward targets set in their NDCs.

In short, I argue that what Ecosystem Marketplace (2021) articulated in the REDD+ context also applies to other forest sector NbSs:

> A jurisdictional REDD+ program is a government-led program to address the drivers of deforestation and forest degradation and to enhance forest C stock over a large jurisdiction of forest cover across subnational and national scales. Jurisdictional programs are different from REDD+ projects, which cover a relatively small area, carry out activities (as opposed to policies and regulations) to address local drivers of deforestation and degradation, and are often implemented by civil society organizations or private companies.

## Advantages of jurisdictional approaches

Now, let me examine the comparative advantages of jurisdictional approaches more closely, as reflected in the compatibility with the Paris Agreement's carbon accounting requirements, the essential capacity of measurement and MRV, and the moderation of entry barriers and costs. These elements are inherently connected, and it is crucial to assess them carefully, provided that carbon offsetting and crediting from the terrestrial ecosystems pose challenges for meeting accounting and other requirements.

## Compatibility with accounting requirements

Low confidence in REDD+ and other projects and offsetting credits hinges on the difficulty of assuring permanence and additionality. One common method to address these concerns is to install buffer reserves—each project contributes a share of the credits achieved, which works as risk insurance (Bossio et al. 2020). Risk comes from intentional events (e.g., land degradation or conversion) or unintentional events (storms, flooding, fire, and the like). Once these events cause sink reversals or stock losses, credits held in the buffer account equivalent to the amount of reversal will be released and permanently canceled. Most voluntary carbon market standards operate with such buffer reserves (Galik et al. 2014).

In Australia, for instance, carbon farming associated with the government's land-based strategies for climate mitigation follows a mix of restrictions that combine buffer reserves with discount elements: farmers that would receive a certain number of credits in a 100-year permanence scenario would receive 20% fewer credits if they committed to 25-year stable conditions only; the discount comes on top of the general 5% buffer amount (DeFries et al. 2022). Similarly, the California emission trading scheme has a formula for quantifying the offset project risks, which fall in the range of ~9–16% (CARB 2021). But these are substantial impediments; coupled with the low market prices, they have caused the incentives for generating forest carbon offsetting credits to remain negligible.

A large, regional-scale carbon stock can buffer events that might occur at the individual project level (Wang et al. 2021). That is, once an emission reduction/removal action is covered under the NDC, the non-permanence issue in a specific project is resolved in a higher-level accounting scheme. DeFries et al. (2022) further illuminated that jurisdictional approaches potentially lower transaction costs and reduce problems of additionality and leakage. At a project level, additionality is compromised when voluntary participation in response to an economic incentive is biased toward land users who already intended to change practices. This "adverse selection" problem diminishes as actors are grouped within larger jurisdictions, and jurisdiction-wide support for low-carbon land management enables other land users to participate. Consistent standards and dynamic adjustments in a structure predefined by crediting agencies could help reduce the problem of adverse selection at a jurisdictional level (van der Gaast et al. 2017).

## Accountability of jurisdictional governance

Furthermore, it is posited that adopting jurisdictional approaches is a recognition of the vital role that governments at multiple levels must play in any realistic attempt to reduce emissions and enhance removals (Boyd et al. 2018). Government and non-governmental organizations, including voluntary market participants, should thus work together within a given jurisdiction and complement the authority in delivering its commitments. Jurisdictional approaches also offer opportunities for experimentation and policy innovation, including partnerships with supply chain actors and indigenous communities (Oldfield et al. 2022). Thus, it is central for our policy dialogue to adequately reflect this fundamental reality and, if possible, forest carbon projects should be nested within a jurisdictional program (Seymour and Busch 2016). Jurisdictional approaches rely on administration by a public or private entity that would set criteria for the generation and independent verification of credits, to which protocols and project developers must adhere.

Also, jurisdictional funding represents an important turning point in carbon financing because until recently, most forest sector results-based finance has involved market-based funding, while the designated public-sector funding has been funneled to REDD readiness efforts (Ecosystem Marketplace 2021). Jurisdictional funding has also opened up new opportunities for "nesting" and "hybridizing" market-based, public, and other private alternatives; so, it offers distinct advantages that are expected to accelerate funding of forest sector NbSs (Seymour 2020). For example, unlike project developers, governments have the authority to enforce policy and control land use change, while the private sector can serve as a source of immediate results-based payments.

## *Indispensable capabilities*

Although REDD+ was conceived as a mechanism at the national scale, many of the early actions have been through individual projects, giving rise to local challenges such as meeting accounting requirements, assuring buyer confidence, and securing local rights. Meanwhile, countries are mandated to develop supportive systems to operationalize REDD+, including a National REDD+ Strategy or Action Plan, a National Forest Monitoring System, Forest Reference Emission Levels or Forest Reference Level, and a Safeguards Information System (FAO 2016). These are daunting tasks, which can only be fulfilled by competent national and/or subnational authorities.

To illustrate, let us look at the status of the National Forest Inventory (NFI) system. According to FAO (2020), the NFI system in many countries was substantially improved in support of the Global Forest Assessment Report (FRA) 2020. Compared to that underlying FRA 2015, many produced new data and enhanced their monitoring and reporting. Nevertheless, for a number of other variables included in FRA 2020, the data coverage and quality remain poor, and most countries are still reporting on biomass and carbon using default factors. Regional, local forest inventory systems are uncommon and, if they are in existence, their reliability and coverage may be questionable, diminishing the likelihood of scaling up REDD+ and other initiatives. Besides, assessing the reference level and additionality using national apparatus and data will alleviate a country's burden and increase its consistency in dealing with international agencies in a unified way (Mertz et al. 2018).

Oldfield et al. (2022) also explained that monitoring change in GHGs across a region has several benefits. The greater accuracy of models at regional scales resolves uncertainty generated at the field or project level. A network of sites would allow for verification of BAU or reference level and improved management practice scenarios specific to stratified zones within each region, providing a reduced suite of measurements to allow for true out-of-sample model validation. Also, a higher aggregate-level inventory of land use and economic practices can provide data that would allow for more transparent and consistent accounting and thus save on transaction costs and allow greater revenues to be passed to providers. Pooled capabilities of measurement and MRV at a jurisdictional level could facilitate the participation of smallholders and promote desirable land management practices.

*Entry barriers and transaction costs*

Kaarakka et al. (2021) observed that existing forestry offset projects across the contiguous U.S. have an average size of 22,240 acres. In contrast, family forest landowners in the U.S. own on average 67.2 acres of forestland. Likewise, small forest owners do not meet the requirements for offset project permanence, accounting, and monitoring, and are struggling with the rising costs of both the carbon inventory and accounting, as at least 1500–5000 acres are needed to balance the transaction cost of participating in the offset market. The project development and verification costs are high as well—more than $100,000 and $45,000 respectively, pushing participation out of the reach of many small owners. Similarly, Gu et al. (2019) documented the expenses of project development and overhead fees for two counties in Zhejiang, China—1.9 million CNY for the Suichang Carbon Project (8,636ha) and 1.8 million CNY for the Anji Carbon Project (1,426ha). Included are fees for the design and documentation of the management plan, the review and approval, the monitoring and reporting, the certification, and the overhead. As a result, the county governments had to assume these charges.

Unlike individual project-based efforts, jurisdictions or region-wide organizations can foster low-carbon land management through credit and access to inputs for land users to plant trees and improve management, or other practices that require upfront investments. These efforts could be further linked up with enforced regulations within jurisdictions against deforestation, fire, and other disasters. For example, municipality-level initiatives in the Brazilian Amazon successfully reduced deforestation by linking credit to farmers with municipal deforestation rates in executing the federal Plan to Prevent and Control Amazon Deforestation (Parrotta et al. 2022).

Small-scale farmers, constituting 84% of all farmers globally and indigenous communities, are at a disadvantage to participate in the current carbon market, yet they could contribute much to climate mitigation (Ecosystem Marketplace 2021). A carbon market that purchases offsets from land users in jurisdictions that promote low-carbon land management could also help address the existing disincentives to potential suppliers from high transaction costs. McDermott et al. (2022) warned that a further concern is the proliferation of technical needs—especially for monitoring, reporting, and accountability, which have created an industry of intermediaries and consultants. These entities corner a significant portion of forest carbon finance, but their activities do not directly contribute to ER&R. The costs of engaging this level of sophisticated expertise also creates barriers to entry for small landowners and projects that are led by indigenous peoples.

**Policy and practice implications**

Achieving the long-term temperature targets of the Paris Agreement requires forest-based mitigation (IPCC 2022, Grassi et al. 2021). The main objective of this chapter was to deliberate the adequacy and appropriateness of different approaches to governing forest sector climate actions. There are alternative approaches—project-based or jurisdictional—to governing forest sector climate actions. While jurisdictional actions must be ultimately disaggregated into specific projects for implementation, they have some clear advantages. These advantages are reflected in the effectiveness and efficiency of delivering the responsibilities of emission reduction and removal while controlling the entailed risk and cost through better-developed baselines and MRV infrastructure, consistent accounting, wider participation, and strengthened market integrity (DeFries et al. 2022).

It should be pointed out that adopting a jurisdictional approach is not necessarily in conflict with the notion of polycentric governance—a self-organizing governance system comprised of multiple types of actors, decision-making venues, and policy issues of concern, as well as the linkages between these elements, as advocated by Ostrom (2010). Of course, the underpinning of a jurisdictional approach to climate governance has to do with the nature of climate as a global public good or externality (Nordhaus 2019), which calls for a broad, effective, and substantive participation of the whole global community in stabilizing the climate and controlling the temperature rise. The nesting and integration of all stakeholders and participants is also a reflection of the subsidiarity principle (Nordhaus 2021), which means that from the international perspective, it is appropriate to focus its attention on how national jurisdictions carry out the targets specified in their NDCs. At the same time, national (or subnational) authorities should take responsibility for the forest carbon programs and projects to be implemented within their jurisdictions.

Accordingly, every social organization should do its part within a unified framework to save costs and enhance effectiveness of climate actions. Thus, market-based and non-market measures should be subject to the chosen approach(s) to governance, and different components of the system should operate cohesively (Ostrom 2010). In this light, forest sector climate actions should be integrated with other mitigation and adaptation commitments within the NDC; likewise, such organizations as intermediaries or aggregators, verifiers, certifiers, and other market participants should all work together for the shared goal of climate stabilization and sustainable development (UNFCCC 2022). In view of the murky, chaotic, and even speculative behaviors of the voluntary carbon markets (Zadek 2023), it is incumbent upon the national or subnational authorities to develop a more conducive and effective regulatory framework in which a healthy carbon market, especially a voluntary one, can flourish and be trusted and participation in it will be facilitated.

Nonetheless, policymakers, practitioners, and analysts ought to understand the limitations of jurisdictional approaches, such as political turnover, limited public sector capacity, and the lack of broader support and incentives (Boyd et al. 2018, Wunder et al. 2020), all of which can hamper the kind of long-term, sustained attention forest sector initiatives require to succeed. Employing a jurisdictional approach may also face hurdles in places with rampant corruption and weak enforcement (DeFries et al. 2022). Jurisdictional approaches can potentially provide alternative paths to individual projects if participants overcome these hurdles, but coordination across large areas can be difficult. Likewise, leakage could still occur with neighboring jurisdictions and across national boundaries, a problem that is hard to address through the market or other mechanisms. At the same time, internal disaggregation and sharing of commitments can be worked out at the subnational level(s) more efficiently, allowing for reference levels to be adjusted and double counting to be avoided (Schneider et al. 2019).

Another overriding concern for all approaches is the need for fair and equitable investments and benefits sharing (Pachauri et al. 2022, FAO 2016). But land rights and governance are deeply political and involve entrenched power dynamics anyway (McDermott et al. 2022), and it is unrealistic to expect that the market or jurisdiction can easily resolve centuries-old problems of insecure land rights and procedural injustice (DeFries et al. 2022). National REDD+ strategies must combine national coordination and policy cohesion with meaningful local involvement in implementation (Parrotta et al. 2022). As one more step toward the goal of reducing injustice, credit buyers on the carbon market can insist that strong and verified safeguards, principles of informed consent, and fair benefit sharing are enforced. Even when safeguards are put in place, however, they are not a guarantee against unfair and inequitable practices, particularly where governments control forest lands that local people use customarily (Brandão et al. 2020, Yin et al. 2016).

In closing, jurisdictional approaches are better positioned to facilitate scaling up forest sector climate actions, despite their complexity and the need for greater efforts of building capacity and controlling bureaucracy. Our analysis suggests a strong need for reorienting the inquiry into the issues of governing forest sector climate actions from the predominantly disaggregated, project-based paradigm to a more integrated, program-oriented one, such that it will be more congruent with and conducive to delivering the NDC targets (Wang et al. 2021). When integrating REDD+ and other forest sector efforts into NDCs, however, actions in the LULUCF arena have unique characteristics that must be carefully considered. Hopefully, not only can these efforts enhance the NDC ambition and execution, but lessons learned from REDD+ implementation can also help to inform technical guidance, policy dialogue, and practical initiatives related to the broader range of NbSs, contributing to the realization of the full potential of nature to achieve global climate goals (UN REDD Programme 2022). The research community can play a key role in developing the evidence base about problems and successes in implementing forest sector climate actions.

## Note

1 See the Introduction for a definition of REDD+.

## References

Bossio, D.A., Cook-Patton, S.C., Ellis, P.W., Fargione, J., Sanderman, J. et al. 2020. The role of soil carbon in natural climate solutions. *Nature Sustainability* 3 (5): 391–398, https://doi.org/10.1038/s41893-020-0491-z.

Boyd, W., Stickler, C., Duchelle, A.E., Seymour, F., Nepstad, D. et al. 2018. *Jurisdictional Approaches to REDD+ and Low Emissions Development: Progress and Prospects.* Washington, DC: World Resources Institute.

Brandão, F., Piketty, M-G., Poccard-Chapuis, R., Brito, B., Pacheco, P., Garcia, E., Duchelle, A.E., Drigo, I., Peçanha, J.C. 2020. Lessons for Jurisdictional Approaches from Municipal-Level Initiatives to Halt Deforestation in the Brazilian Amazon. *Front. For. Glob. Change* 3: 96. doi:10.3389/ffgc.2020.00096.

California Air Resources Board (CARB). 2021. *California's Compliance of Forest Sector Program.* Released October 27, 2021. Available at: ww2.arb.ca.gov/sites/defa ult/files/2021-10/nc-forest_offorest sectoret_faq_20211027.pdf.

Chagas, T., Galt, H., Lee, D., Neeff, T., Streck, C. 2020. *A Close Look at the Quality of REDD+ Carbon Credits.* Available at: www.climatefocus.com.

DeFries, R., Ahuja, R., Friedman, J., Gordon, D.R., Hamburg, S.P., Kerr, S., Mwangi, J., Nouwen, C., Pandit, N. 2022. Land management can contribute to net zero. *Science* 376 (6598): 1163–1165.

Ecosystem Marketplace. 2021. *State of Forest Carbon Finance 2021*. Washington DC: Forest Trends Association.

Food and Agriculture Organization of the United Nations (FAO). 2020. *Global Forest Resource Assessment Report.* Rome, Italy. https://doi.org/10.4060/ca9825en.

Food and Agriculture Organization of the United Nations (FAO). 2016. *Forestry for a Low -Carbon Future: Integrating Forests and Wood Products in Climate Change Strategies* (FAO Forestry Paper 177). Rome, Italy.

Fischer, R., Hargita, Y., Günter, S. 2016. Insights from the ground level? A content analysis review of multi-national REDD+ studies since 2010. *Forest Policy and Economics* 66: 47–58.

Galik, C., Murray, B.C., Mitchell, S., Cottle, P. 2016. Alternative approaches for addressing non-permanence in carbon projects: an application to afforestation and reforestation under the Clean Development Mechanism. *Mitigation and Adaptation Strategies for Global Change* 21: 101–118.

Grassi, G., Stehfest, E., Rogelj, J., van Vuuren, D., Cescatti, A. et al. 2021. Critical adjustment of land mitigation pathways for assessing countries' climate progress. *Nature Climate Change* 11: 425–434.

Griscom, B.W., Adams, J., Ellis, P.W., Houghton, R.A., Lomax, G., Miteva, D.A., Schlesinger, W.H., Shoch, D., Siikamäki, J.V., Smith, P. et al. 2017. Natural climate solutions. *PNAS* 114 (44): 11645–11650.

Groom, B., Palmer, C., Sileci, L. 2022. Carbon emissions reductions from Indonesia's moratorium on forest concessions are cost-effective yet contribute little to Paris pledges. *PNAS* 119 (5): e2102613119.

Gu, L., Wu, W.G., Jia, W., Zhou, M.J., Xu, L., Zhu, W.Q. 2019. Evaluating the performance of bamboo forests managed for carbon sequestration and other co-benefits in Suichang and Anji, China. *Forest Policy and Economics* 106 (101947).

International Panel on Climate Change (IPCC). 2020. *Climate Change and Land: Summary for Policymakers.* Available at: www.ipcc.ch/srccl/.

International Panel on Climate Change (IPCC). 2022. *Climate Change 2022: Impacts, Adaptation and Vulnerability.* Working Group II Contribution to the Sixth Assessment Report.

Irawana, S., Widiastomo, T., Tacconi, L., Watts, J.D., Steni, B. 2019. Exploring the design of jurisdictional REDD+: The case of Central Kalimantan, Indonesia. *Forest Policy and Economics* 108 (101853).

Kaarakka, L., Cornett, M., Domke, G., Ontl, T., Dee, L.E. 2021. Improved forest management as a natural climate solution: A review. *Ecological Solutions and Evidence*: 7. doi:10.1002/2688-8319.12090.

McDermott, C., Vira, B., Walcott, J., Brockhaus, M., Harris, M., Kumeh, E.M., de Mendonça Gueiros, M. 2022. *The Evolving Governance of REDD+* (Chapter 2 of Parrotta et al. 2022).

Mehling, M.A., G.E. Medcalf, R.N. Stavins. 2018. Linking climate policies to advance global mitigation. *Science* 359 (6379): 997–998.

Mertz, O., Grogan, K., Pflugmacher, D. et al. 2018. Uncertainty in establishing forest reference levels and predicting future forest-based carbon stocks for REDD+. *Journal of Land Use Science* 13: 1–15.

Midkiff, D., Yin, R.S., Zhang, H. 2022. *Carbon Finance and Funding for Forest Sector Climate Solutions: A Synthesis of Principles, Mechanisms, and Developments* (MSU working paper).

Murray, B.C., Sohngen, B., Ross, M.T. 2006. Economic consequences of consideration of permanence, leakage and additionality for soil carbon sequestration projects. *Climatic Change* 80: 127–143.

Nepstad, D.C., Boyd, W., Stickler, C.M., Bezerra, T., Azevedo, A.A. 2013. Responding to climate change and the global land crisis: REDD+, market transformation and low-emissions rural development. *Philosophical Transactions of the Royal Society B* 368 (1619): 20120167.

Nordhaus, W. 2019. Climate Change: The Ultimate Challenge for Economics. *American Economic Review* 109: 1991–2004. Available at: https://doi.org/10.1257/aer.109.6.1991.

Nordhaus, W. 2021. *The Spirit of Green.* New Jersey: Princeton University Press.

Oldfield, E.E., Eagle, A.J., Rubin, R.L., Rudek, J., Sanderman, J., Gordon, D.R. 2022. Crediting agricultural soil carbon sequestration. *Science* 375 (6586): 1222–1225.

Ostrom, E. 2010. Beyond markets and states: polycentric governance of complex economic systems. *American Economic Review* 100: 641–672.

Pachauri, S., Pelz, S., Bertram, C., Kreibiehl, S., Rao, N.D., Sokona, Y., Riahi, K. 2022. Fairness considerations in global mitigation investments. *Science* 378: 1057–1059.

Parrotta, J., Mansourian, S., Wildburger C., Grima, N. (eds). 2022. *Forests, Climate, Biodiversity and People: Assessing a Decade of REDD+*. IUFRO World Series Volume 40. Vienna.

Republic of Indonesia. 2022. *Enhanced Nationally Determined Contribution Submitted to UNFCCC Secretariat*. Available at: https://unfccc.int/reports.

Roberts, J.T., Weikmans, R., Robinson, S., Ciplet, D., Khan, M., Falzon, D. 2021. Rebooting a failed promise of climate finance. *Nature Climate Change* 11: 180–182.

Roe, S., Streck, C., Obersteiner, M., Frank, S., Griscom, B., Drouet, L. et al. 2019. Contribution of the land sector to a 1.5°C world. *Nature Climate Change* 9: 817–828. doi:10.1038/s41558-019-0591-9.

Roopsind, A., Sohngen, B., Brandt, J. 2019. Evidence that a national REDD+ program reduces tree cover loss and carbon emissions in a high forest cover, low deforestation country. *PNAS* 116 (49): 24492–24499.

Schneider, L., Duan, M.S., Stavins, R., Kizzier, K., Broekhoff, D., Jotzo, F., Winkler, H., Lazarus, M., Howard, A., Hood, C. 2019. Double counting and the Paris Agreement rulebook. *Science* 366 (6462): 180–183.

Seymour, F. 2020. *INSIDER: 4 Reasons Why a Jurisdictional Approach for REDD+ Crediting I Superior to a Project-Based Approach*. World Resources Institute, May.

Seymour F., Aurora, L., Arif, J. 2020. The Jurisdictional Approach in Indonesia: Incentives, Actions, and Facilitating Connections. *Front. For. Glob. Change* 3: 503326. doi:10.3389/ffgc.2020.503326.

Seymour, F., Busch, J. 2016. *Why Forests? Why Now? The Science, Economics, and Politics of Tropical Forests and Climate Change*. Washington, DC: Brookings Institution Press.

Thamo, T., Pannell, D.J. 2016. Challenges in developing effective policy for soil carbon sequestration: perspectives on additionality, leakage, and permanence. *Climate Policy* 16: 973–992.

United Nations Framework Convention on Climate Change (UNFCCC) Secretariat. 2013. *Afforestation and Reforestation Projects under the Clean Development Mechanism: A Reference Manual Reference Manual for the Enhanced Transparency Framework under the Paris Agreement*. Bonn, Germany.

United Nations Framework Convention on Climate Change (UNFCCC) Secretariat. 2015. *Synthesis Report on the Aggregate Effect of the Intended Nationally Determined Contributions*. Available at: https://unfccc.int/resource/docs/2015/cop21/eng/07.pdf.

United Nations Framework Convention on Climate Change (UNFCCC) Secretariat. 2020. *Reference Manual for the Enhanced Transparency Framework under the Paris Agreement*. Bonn, Germany.

UN REDD Programme. 2022. *Linking REDD+, the Paris Agreement, Nationally Determined Conttributions and the Sustainable Development Goals: Realizing the Potential of Forests for NDC Enhancement and Implementation*. Available at: www.un-redd.org/sites/default/files/2022-03/NDC%20Final.pdf.

Van der Gaast, W., Sikkema, R., Vohrer, M. 2018. The contribution of forest carbon credit projects to addressing the climate change challenge. *Climate Policy* 18 (1): 42–48.

Von Essen, M., Lambin, E.F. 2021. Jurisdictional approaches to sustainable resource use. *Frontiers in Ecology and the Environment* 19 (3): 159–167. doi:10.1002/fee.2299.

Wang, Y., Li, L., Yin, R., 2021. A primer on forest carbon policy and economics under the Paris Agreement: Part I. *Forest Policy and Economics* 132 (102595).

West, T.A.P., Borner, J., Sills, E.O., Kontoleon, A. 2020. Overstated carbon emission reductions from voluntary REDD+ projects in the Brazilian Amazon. *PNAS* 117 (39): 24188–24194.

World Bank. 2021. *State and Trends of Carbon Pricing*. Washington, DC.

Wunder, S., Duchelle, A.E., Sassi, C., Sills, E.O., Simonet, G., Sunderlin, W.D. 2020. REDD+ in theory and practice: How lessons from local projects can inform jurisdictional approaches. *Front. For. Glob. Change* 3: 11. doi:10.3389/ffgc.2020.00011.

Yin, R.S., Zulu, L., Qi, J.G., Freudenberger, M., Sommerville, M. 2016. Empirical linkages between devolved tenure systems and forest conditions: Primary evidence. *Forest Policy and Economics* 79: 277–285.

Zadek, S. 2023. Trouble with carbon markets. *Project Syndicate* (March 1).

# 3 Forest carbon accounting and assessment

Alternative perspectives and methodologies

## Introduction

Forest ecosystems play a considerable role in the global carbon cycle (IPCC 2020), and the Paris Agreement has included a wide range of forest sector actions in the international efforts of climate change mitigation (Roe et al. 2019). Moreover,

> Agriculture, Forestry, and Other Land Use (AFOLU) is unique among the sectors … since the mitigation potential is derived from both an enhancement of removals of greenhouse gases, as well as a reduction of emissions through management of land and livestock.
> 
> (Smith, P. et al. 2014)

As a large part of the "nature-based solutions" (Griscom et al. 2017), emission reduction and removal (ER&R) of greenhouse gases (GHG) in the forest sector are commonly analyzed for the purposes of carbon accounting and assessment, to which there exist alternative approaches (Ohrel 2019, Nabuurs et al. 2010). Overall, these approaches constitute direct and indirect inventory-based methodologies, as deliberated below. It is beneficial and timely to profile them and examine their basic formulation and functionality, as well as comparative strengths, which is what I do in this chapter.

Carbon accounting is the process of measuring, calculating, analyzing, and reporting an organization's ER&R (UNFCCC 2020). Jurisdictions, businesses, and other entities use accounting to manage their carbon footprint; likewise, the UN Climate Secretariat and the Conference of Parties employ carbon accounting to track a Party's progress relative to its nationally determined contribution (NDC) and to conduct global stocktaking of the announced NDCs in combination against the Paris Agreement's climate targets. Carbon assessment is the process of evaluating and predicting the potential effects of socioeconomic scenarios or policy options to inform decision-making (Ohrel 2019, Lan and Yin 2017). There exist many carbon policy options within the forest sector, and they are assessed in terms of their effectiveness, efficiency, and equity (Pagiola et al. 2004, Roe et al. 2019). For

DOI: 10.4324/9781003436652-4

example, we are interested in what the effects of carrying out the REDD+[1] initiative on global ER&R would be if payments could be made at different price levels, say, $20 or $50 per ton of carbon dioxide ($CO_2$). Similarly, it is imperative to determine what a potential role the forest sector can play in a country's ER&R endeavors, including achieving the goal of carbon neutrality or net zero, under different circumstances, until 2030 or the mid-century.

The effects of socioeconomic scenarios or policy options can be evaluated at the project or program level as well as for the whole forest sector (Ohrel 2019, Lan and Yin 2017). Likewise, they can be evaluated *ex ante* or *ex post*. Here, my focus is on sector-wide assessment both *ex ante* and *ex post*. I will consider evaluating the *ex post* impacts of forest carbon projects and programs in Chapter 9. While carbon accounting is focused on how a particular organization is doing in fulfilling its voluntary and/or compulsory commitments to ER&R, carbon assessment is geared toward evaluating the sectoral effects of a particular scenario or option. However, carbon accounting and assessment are not necessarily mutually exclusive. Just like accounting can be used for projecting the potential effects of a relevant policy scenario, assessment can be used for analyzing an entity's ER&R progress. Therefore, it is crucial for researchers and practitioners in economics, ecology, and other disciplines to broaden their analytic scope and toolkit by embracing both policy assessment and carbon accounting. Indeed, at the exact time Parties to the UNFCCC begin to fulfill their existing commitments and raise their targets to higher levels, the need for forest sector carbon accounting and assessment is not only enormous but also urgent (Roe et al. 2019, Austin et al. 2020).

Forest inventory data are directly used in carbon accounting for determining an entity's ER&R (Nabuurs et al. 2010, Smith et al. 2018), whereas inventory data are often linked up with some kind of modeling system of timber, or wood products, markets and thus used indirectly for carbon assessment (Buongiorno et al. 2003, Sohngen et al. 1999, Adams and Haynes 1996). As such, the existing literature on forest carbon accounting and assessment can be classified into two main strands—one featuring the direct inventory-based approach and the other an indirect inventory-based approach. Also, these two approaches can be adopted at different levels of aggregation—international, national, regional, or local. Ohrel (2019) reviewed the literature that features the indirect inventory-based approach, but it did not get much into the other strand of the literature that features the direct inventory-based approach. Moreover, it seems that some analysts and practitioners have not yet fully recognized the eminence, as well as imminence, of carbon accounting (United Nations 2022), which is not conducive to advancing the global climate agenda and must be changed.

Van Kooten (2018) nicely stated that "economic agents must know the rules of the carbon game before making forestry decisions, and these rules are ultimately established by the political authority." What are "the rules of the carbon game"? We know that the architecture of the Paris Agreement centers around the NDCs. Article 4.13 of the Paris Agreement stipulates that

Each Party shall regularly provide the following information: (a) A national inventory report of anthropogenic emissions by sources and removals by sinks of greenhouse gases, prepared using good practice methodologies accepted by the IPCC and agreed upon by the COP ... and (b) Information necessary to track progress made in implementing and achieving its NDC.

In preparing and communicating an NDC, as noted in Chapter 1, a Party begins by articulating its concrete actions of ER&R and selecting indicators for those actions; once the indicators are selected, inventory data should be gathered on each of them, including reference information and any updated statistics from previous years. Thus, Parties need to select their own methodologies and metrics in reference to the IPCC Guidelines.

Hence, we may summarize the essential PA stipulations and requirements as follows:

1. a jurisdictional approach to the governance of climate mitigation actions, including those of the forest sector, is predicated on the NDC architecture;
2. the NDC is made up of concrete targets of ER&R, which require a Party to select necessary indicators for the chosen actions and gather the inventory data on each of them;
3. accounting and assessment should be conducted with respect to such principles as additionality, permanence, and leakage avoidance; and
4. just like an NDC can be disaggregated into sector-specific actions and targets, it can also be decomposed into subnational actions and targets.

It should now become clear that inventory information at a jurisdictional level is crucial for conducting stock-taking, progress-tracking, and tasks of MRV, or monitoring, reporting, and verification (UNFCCC 2020).

In this chapter, I begin with an overview of the basic formulation and functionality of the two alternative approaches before delving into their comparative strengths. Then, I use four case studies from Canada, Russia, and Europe, as well as the whole world, to illustrate how the different approaches are applied and what insights they have generated. Finally, I make a few closing remarks on the research gaps and future directions. I hope that my efforts of literature review and synthesis will help policy-makers, practitioners, and analysts alike put the related issues in perspective and facilitate more effective deployment of the different approaches to carbon accounting and assessment along the way of achieving the ER&R targets of various jurisdictional and other organizations. In writing this chapter, I referenced a large body of literature, drawing upon information and insight, particularly from Wang et al. (2021), Ohrel (2019), Smith et al. (2018), and Nabuurs et al. (2010).

## Alternative methodologies

It is useful to reiterate the forest carbon pools and thus the basic data needs before I describe the alternative approaches. As set out in Chapter 1, forest sector carbon stock is made up of carbon sequestered by forest ecosystems (FESs) and stored in harvest wood products (HWPs) (IPCC 2006). Forest carbon fluxes thus pertain to and are reflected in the changes of these carbon pools (or stocks) over time. So, the specific information required for accounting and assessment includes inventories of forest biomass (above and below ground), forest soil, and in-service wood products (IPCC 2006).

In general, forest biomass has not been historically measured as part of a typical national forest inventory (NFI) system, which was and still is focused on the stock volume, or stem-wood as used by Nabuurs et al. (2010), of the regular forests[2] (Zeng et al. 2015, Liu and Yin 2012). Nevertheless, forest biomass volume can be derived from the stock volume by establishing a relationship between them and then carbon stock can be estimated from biomass stock using a carbon conversion factor (Piao et al. 2009, FAO 2020). Of course, there exist more sophisticated inventory-based methods than conversions of the stock volume from NFI data. As an example, Nabuurs et al. (2010) introduced several studies conducted in Europe, using such methods as whole ecosystem carbon balances, remote-sensing based techniques, micro-meteorological techniques and their upscaling, and large-scale ecosystem models.

A data gap is often found in forest soil, which has rarely been systematically measured (Kaarakka et al. 2021, Nabuurs et al. 2010). China is a case in point. Piao et al. (2022) stated that periodical and standardized inventory of soil carbon stock is still lacking at a national scale, especially in the central and western regions of the country. Without all the necessary inventory information, though, it would be impractical for a Party to pursue a comprehensive forest carbon accounting or assessment. Meanwhile, carbon stored in HWPs can be estimated using the annual data on production, trade, and consumption of forest products provided by the United Nations FAO (www.fao.org/forestry/statistics/84922/en/), or data from another source, coupled with one of the methods recommended by the IPCC (2006, 2019).

According to Ohrel (2019), included in the alternative approaches are "historic reference and extrapolations, ecological models, forest sector economic models, integrated assessment models, and meta-analysis." Here, I offer a different perspective; in my view, there exist two basic types of approaches—direct and indirect inventory-based, as already noted. The former, the same as the "historic reference and extrapolations" of inventory data described by Ohrel (2019), should be straightforward to understand. It makes future projections of and evaluate policy effects on forest conditions using historical data of growth, yield, area, and so on in combination with necessary assumptions and/or modules of ecological, land-use change, and other modeling components.

Obviously, omitting economics may lead to biased ER&R estimates. In most parts of the world, however, a market model of timber, or wood products, is simply not available, whereas inventory data are (FAO 2020). In this situation, only the direct inventory-based approach is feasible. It can also be argued that, after all, because we deal with $CO_2$ (and other GHGs) in the context of climate change mitigation, not timber per se, a timber (or wood products) market modeling system has to work backward to the inventory condition before deriving the carbon stock and/or flux associated with the stock volume, not to mention other pools like forest soil and HWP storage. In fact, an NFI system of decent quality has not been established yet in certain countries (FAO 2020), making the use of a direct inventory-based approach challenging.

For a market-based modeling system, the same as Ohrel's (2019) "forest sector economic models," forest inventory is included as a driver of the timber supply, whereas the demand for timber is derived from the demand for wood products (Sohngen and Mendelsohn 2003, Austin et al. 2020). In other words, the core of a market model is made up of a set of wood products supply and demand equations subject to the maximization of social welfare—a sum of consumer and producer surpluses, with a module of forest inventory dynamics linked with the market model (Wear and Coulston 2019). Once developed, such a modeling system can be used to make simulations and projections of the inventory and other variables of concern.

The so-called "ecological models" are often used to capture the inventory response to environmental changes, such as carbon fertilization, nitrogen deposition, and/or forest fire (Grassi et al. 2021, Mendelsohn and Sohngen 2019), and they are members of the family of direct inventory models (Nabuurs et al. 2010). Finally, "integrated assessment models," or IAMs, are typically modeling systems that integrate models of economy, energy, and/or ecology to assess potential policy outcomes and tradeoffs, allowing insights not possible in single sector models (Nordhaus 2021). Often, IAMs have highly aggregated sector representation (Rennert et al. 2021), which does not offer detailed evaluation of specific forest ecosystem, and market interactions and related GHG outcomes. On the other hand, "meta-analysis" is commonly deployed to extract findings and other useful information from existing studies of possibly different approaches (Maziarz 2022, Roe et al. 2019), which may include direct and indirect inventory-based models, and even IAMs.

Of course, both direct and indirect inventory-based approaches need to consider land-use and/or management technology changes in making future projections (Johnston and Radeloff 2019, Smith et al. 2014). Depending on the degree of sophistication, these components may simply be assumed scenarios or carefully formulated shared socioeconomic pathways (SSPs) (e.g., IPCC 2022, Daigneault et al. 2019). More formally, though, modular components are often integrated with the market model of an indirect inventory-based approach (and, indeed, the direct inventory-based approach as well) to capture the fundamental interaction of supply and demand forces and thus embedded in a large, complex modeling system (Adams and Haynes 1996, Sohngen et al. 1999, Buongiorno et al. 2003).

Furthermore, as deliberated in Chapter 1, different approaches can lead to discrepancies between the gross change of forest carbon stock and the counterpart induced by anthropogenic forces because they may not capture the indirect effects resulting from increase in atmospheric $CO_2$ concentrations and other factors. More specifically, a large fraction of land ER&R resulting from indirect human activities are considered 'natural' in IAMs but 'human-caused' in inventories (Grassi et al. 2021). As a result, there has been a persistent gap in the carbon cycle literature between inventory-based estimation and IAMs. By combining these indirect emissions and removals from dynamic global vegetation models with the direct emissions from the IAMs, Grassi et al. (2021) have convincingly eliminated most of the discrepancy. For now, we can determine whether the forest biomass and carbon dynamics are driven by human actions and whether we can conduct carbon accounting and policy assessment based directly on the NFI information and the delineation provided therein.

A related issue is the need for determining to what extent the forest biomass change is induced by anthropogenic or natural forces like fire, carbon fertilization, and/or nitrogen deposition. A separate ecological model is often formulated to capture these effects (Mendelsohn and Sohngen 2019, Nabuurs et al. 2010). It is in this context that I believe the use of ecological modeling is only part of the direct inventory-based approach, albeit in response to a set of specific factors and using different techniques, like biogeochemical simulators and vegetation or process models. It is useful to consider future biophysical potential of forest outcomes when an analysis focuses on physiological parameters such as ecosystem dynamics and climate impacts (IPCC 2020, Favero et al. 2020).

To be sure, these alternative approaches can be and have been effectively combined to make future projections and address policy issues. Also, they have their own limitations at different levels of aggregation. In general, most countries have inventory data available at the national and subnational levels, allowing them to conduct direct inventory-based accounting and analysis. For many countries, however, they simply do not possess a well-developed wood products market and thus a market modeling system. Even for those countries or regions that possess such a system, a counterpart may not be available at the subnational or local level, forcing the analyst to take a direct inventory-based approach (Hou and Yin 2022).

## Comparative strengths of different approaches

It is true that a market modeling system incorporates the underlying economic mechanisms as reflected in the interaction of supply- and demand-side forces, which captures the basic notion of a market economy. As stated by Ohrel (2019), forest sector economic models use resource and other data to simulate future scenarios of market potential or different degrees of competitive market potential, depending on the modeling framework and scope. The ability of this approach to isolate effects per variation of different parameters offers clarity regarding key features of interest in the forest sector (e.g., analysis of different management regimes and/or species mixes) and allows for better characterization of different ecological and economic drivers of forest GHG fluxes (Sohngen and Mendelsohn 2007, Baker et al. 2017). Favaro et al. (2020) further illustrated that one of the benefits of such a sectoral approach is that they include multiple sequestration options and thus capture the interactions across various subsystems for sequestering carbon.

Johnston et al. (2019) explored the use of spatial equilibrium and trade modeling to project future carbon fluxes in forests for 180 countries. They embedded carbon stocks and fluxes into the Global Forest Products Model, or GFPM (Buongiorno et al. 2003), to track fluctuations in carbon stocks in response to market perturbations. They relied on the FAOSTAT database from the FAO to develop econometric estimates of demand and supply functions for 14 wood product groups across all the countries, which were then used to project future equilibrium outcomes. Land use change entered the GFPM through changes to forest area, which was assumed to be a function of evolving demographics and economic growth, and an environmental Kuznets curve relationship associates changes in income per capita to the annual growth rate of forest area (Johnston and Radeloff 2019). Country-specific demographics and economic changes in the future were adapted from the relevant, publicly accessible SSPs.

As one of the few indirect inventory-based studies from Europe, Guo et al. (2019) analyzed the impacts of an increasing bioenergy production on timber harvest and forest growing stock and on the carbon balance of forests using a partial equilibrium model of the Swedish forest sector. Their results suggested that when compared with the base scenario where the current use pattern of forests continues, increased bioenergy production would lead to a 10–14 million $m^3$ increase in the total harvest, depending on the extraction rate of forest residues. Increasing the use of forest residues would reduce the harvest and leave more room for accumulation of the forest stock in the early years, while the stock accumulation would be partially offset by the increased timber harvest in the long run, and increasing bioenergy production could have a negative impact on the carbon balance primarily due to a net loss of carbon stored in forests. Overall, the joint contribution of forest-based mitigation could be significant—equivalent to or higher than 65% of the country's annual GHG emissions.

Nonetheless, forests can provide multiple types of ecosystem services other than timber or wood products (MA 2005), and many of these services may not be provided through the market, given their nature of public good or externality and the lack of developed markets. Likewise, a lot of the ecosystem restoration and reforestation, afforestation, and forest management activities are not solely driven by market forces (MA 2005); rather, they are pushed and pulled by governmental and/or non-governmental organizations

(Nordhaus 2021). As such, it may not be realistic to attempt to capture the provision of these services with a single market modeling system. In fact, Sohngen and Mendelsohn (2003) acknowledged that their model does not consider the myriad of goods and nonmarket services that emanate from forests.

> The sequestration programs (outlined in their work) would affect habitat, water flows, and recreation services because it would encourage older managed forests and substantial conservation forests that are not harvested at all. The resulting changes in ecosystems and the value of these flows should also be carefully integrated into the model and decision making.

Thus, a market modeling system may need to be complemented or constrained by other considerations and assumptions, depending on the questions to be addressed.

In any case, it would be complicated to delineate the forest inventory data for different uses, while lumping them up can make the modeling work less accurate. Additional factors, including market barriers and concerns with data quality and coverage, may cause the modeling outcomes to be only rough as well. Moreover, these models rely on assumptions about future environmental as well as macroeconomic and specific forest market conditions, which adds uncertainty to the simulated outcomes. Ohrel (2019) also recognized that forest ecosystem measurements and related GHG flux quantification are particularly challenging, as these ecosystems have significant spatial and temporal variability, species and environmental heterogeneity, and interconnectedness with other ecosystems.

On the other hand, while a direct inventory-based approach may not adequately capture the underlying supply and demand forces driving the resource changes, it can be predicated on reasonable assumptions or scenarios in making future projections. As mentioned, many efforts have been made with carefully built demographic, socioeconomic, and/or environmental scenarios (e.g., Lemprière et al. 2013, Nabuurs et al. 2010). Moreover, a direct inventory-based approach can be parsimonious and perform well. Following an assessment of the assumptions for existing Forest Inventory Projection Models, Wear and Coulston (2019) examined the implications for forest carbon projections. Comparing model results to observations from re-measured forest inventories in the eastern US, they showed that forest carbon projections would be sensitive to non-harvest disturbances, ownership, and stand-origin. In addition, bias can arise when forest carbon stocks are estimated using correlations between average stock density and biomass aggregates. They concluded that current forest inventories provide a dataset of consistently re-measured plot records that would increasingly support a strong empirical foundation for Forest Inventory Projection Models.

Before proceeding, it is worthwhile to summarize the major studies that show the wide-ranging applications of different approaches. A selective list of the representative studies featuring both approaches is given in Table 3.1. Interested readers can refer to the studies listed there.

Table 3.1 A selective list of forest carbon accounting and assessment studies

| Title | Author(s) | Journal |
|---|---|---|
| *Direct inventory-based* | | |
| Climate change mitigation in Canada's forest sector: a spatially explicit case study for two regions. | Smith C. et al. (2018) | Carbon Balance and Management |
| Quantifying the biophysical climate change mitigation potential of Canada's forest sector. | Smith C. et al. (2014) | Biogeosciences |
| Canadian boreal forests and climate change mitigation | Lemprière et al. (2013) | Environmental Reviews |
| Russian forest sequesters substantially more carbon than previously reported. | Schepaschenko et al. (2021) | Scientific Reports |
| How significant a role can China's forest sector play in decarbonizing its economy? | Hou and Yin (2022) | Climate Policy |
| Specifying forest sector models for forest carbon projections | Wear and Coulston (2019) | J. of Forest Econ. |
| Europe's terrestrial biosphere absorbs 7 to 12% of European anthropogenic $CO_2$ emissions | Janssens et al. (2003) | Science |
| First signs of carbon sink saturation in European forest biomass. | Nabuurs et al. (2013) | Nature Climate Change |
| The European carbon balance. Part 3. | Luyssaert et al. (2010) | Global Change Biology |
| *Indirect inventory-based* | | |
| The net carbon emissions from historic land use and land use change | Mendelsohn and Sohngen (2019) | J. of Forest Econ. |
| Global mitigation potential of carbon stored in harvested wood products | Johnston and Radeloff (2019) | PNAS |
| An optimal control model of forest carbon sequestration | Sohngen and Mendelsohn (2003) | American J. of Agri. Econ. |
| Forests: Carbon sequestration, biomass energy, or both? | Favero et al. (2020) | Sci. Adv. |
| Assessing the long-term interactions of climate change and timber markets on forest land and carbon storage | Favero et al. (2021) | Environ. Res. Lett |
| The economic costs of planting, preserving, and managing the world's forests to mitigate climate change | Austin et al. (2020) | Nature Communications |
| Developing detailed shared socioeconomic pathway (SSP) narratives for the global forest sector | Daigneault et al. (2019) | J. of Forest Econ. |
| Impacts of increasing bioenergy production on timber harvest and carbon emissions | Guo et al. (2019) | J. of Forest Econ. |

| Title | Author(s) | Journal |
|---|---|---|
| *Indirect inventory-based* | | |
| From source to sink: Past changes and model projections of carbon sequestration in the global forest sector | Johnston et al. (2019) | J. of Forest Econ. |
| Importance of cross-sector interactions when projecting forest carbon across alternative socioeconomic futures | Jones et al. (2019) | J. of Forest Econ. |

Let me close this section with the following quote from Ohrel (2019, p. 194):

> Each approach ... offers different perspectives on forest resources as well as different strengths and weaknesses, and they each can be useful in different research applications. However, with the understanding that nothing is constant in the forest sector ... decisionmakers ultimately need to understand as much as possible about the potential future conditions faced by forested lands and those that manage them. To the extent possible, forming future policies by evaluating policy designs from only one discipline and only looking to the past as a future guide should be avoided.

## Case studies

This section presents four case studies to further illustrate the potential uses and strengths of different approaches. The first case highlights the global forest transition, which was analyzed by Mendelsohn and Sohngen (2019) using their Global Timber Model (GTM). The other three cases feature applications of different methods and techniques of a direct inventory-based approach for the estimation and projections of forest inventory and carbon stock by scientists from Canada (Smith et al. 2018), Russia (Schepaschenko et al. 2021), and the European Union (Janssens et al. 2003, Nabuurs et al. 2013).

### *Setting the global forest history straight*

Mendelsohn and Sohngen (2019) is an excellent example to demonstrate the market modeling approach and its power of policy analysis. Houghton (2008) and Houghton et al. (2012) reported that deforestation had caused a net loss of 484 billion tons of carbon dioxide (Gt$CO_2$) since 1900, which is about one third of all man-made emissions. While these studies considered natural regeneration on forestland, they did not consider the substantial private and public investments in fire control, plantation establishment, afforestation and reforestation, and forest management after 1950, as well as the effect of carbon fertilization on a younger forest. Consequently, those studies failed to recognize the forest sector transformation from being a nonrenewable mining activity to a renewable activity over the last century. The younger age structure made global forests a more active carbon sink. With the renewable stock, the forestry investments offset part of the removals and thus maintained the underlying stock.

The question is to what extent the combination of forest investments and carbon fertilization may have helped forests more than regain the original carbon lost from deforestation. Using data on local growth and biomass in 250 forests around the world, the GTM reproduced the historic timber removal rates and management back to 1900. How forests responded to the changing climate and $CO_2$ concentrations was captured by an ecological modular model, which assumed that all these factors would have increased the productivity of forests on average by 0.08% per year from 1900–1950, rising to 0.4% per year by 2010.

Mendelsohn and Sohngen then compared the outcome of a deforestation scenario with subsequent forest management with what would have happened if the natural forest in 1900 had not been removed. Deforestation plus forest management indicated current forests could hold about 94 $GtCO_2$ more than they did in 1900. However, natural forests would have held an additional 186 $GtCO_2$. Human activities have thus caused about 92 $GtCO_2$ of net emissions since 1900. Even back to the 1990s, mature timber was not necessarily the primary source of timber in global markets, which was largely coming from secondary forests that had either regrown naturally or had been replanted. By 2010, there were 290 million ha of moderately managed forestland and 142 million ha of intensively managed plantations. These forests are growing at a faster rate and have more carbon per ha than naturally regenerated forests.

Mendelsohn and Sohngen further argued that "Even if the role of land use and land use change in historic carbon emissions has been relatively small, forests can still be an important part of the solution to curbing global warming." The world's forests can readily sequester more carbon than they do now if governments provide an incentive for forest owners to store carbon. As the incentive grows, more carbon can be sequestered by forests. For instance, a $CO_2$ price of $17/ton would lead to an additional 147 $GtCO_2$, and a price of $50/ton would lead to an additional 367 $GtCO_2$ stored in forests (Sohngen and Mendelsohn 2003). Austin et al. (2020) provided an updated assessment of the forest sector's potential of carbon sequestration and storage in response to different levels of carbon price, reporting that for forests to contribute ~10% of mitigation needed to limit global warming to 1.5°C, carbon prices will need to reach $281/$tCO_2$. The Mendelsohn and Sohngen study has immense implications to future climate policy. Of course, at the highest aggregate (global) level, there is no need to consider some of the Paris Agreement requirements for forest sector ER&R (e.g., leakage avoidance and jurisdictional governance), while others are easily met (e.g., additionality and permanence).

## Carbon accounting and assessment

*Canada*

Scientists at Canadian Forest Service (CFS) have taken a direct inventory-based, systems approach to measuring the carbon fluxes associated with the interaction between human activities (planting, fertilizing, thinning, harvesting) and the forest ecosystem dynamics, including wildfire and pests. This approach considers carbon stored in long-lived product pools and $CO_2$ emissions avoided when wood replaces steel and cement in construction and/or wood biomass replaces fossil fuels in energy production. As a representative study, Smith et al. (2018) assessed the potential of the forest sector to mitigate GHG emissions by changes in management practices and wood use for two regions within Canada's managed forest from 2018 to 2050.

FES carbon dynamics were estimated using the Carbon Budget Model of the Canadian Forest Sector. Meanwhile, carbon transferred from forest ecosystems to HWPs and bioenergy were tracked through manufacturing, use/export, and end-of-life use by the Carbon Budget Modeling Framework for Harvested Wood Products. Based on a spatially explicit forest inventory with 16 ha pixels, Smith et al. (2018) examined mitigation scenarios of increased harvesting efficiency, residue management for bioenergy, reduced harvest, reduced slash-burning, and more longer-lived wood products. The reason for the spatially explicit approach at this coarse resolution was to quantify transportation distances associated with delivering harvest residues for heat and/or electricity production for local communities.

Their results demonstrated large differences among the alternative scenarios and from alternative assumptions about substitution benefits for fossil fuel-based energy and products. Combining forest management activities with a wood-use scenario that generated more longer-lived products had the highest mitigation potential. The use of harvest residues to meet local energy demands in place of burning fossil fuels was found to be an effective scenario to reduce GHG emissions, along with scenarios that increased the utilization level for harvest and increased the longevity of wood products. Substitution benefits from avoiding fossil fuels or emissions-intensive products were dependent on local circumstances for energy demand and fuel mix and the assumed wood use for products. As projected future demand for biomass use in national GHG mitigation strategies could exceed sustainable biomass supply, analyses such as this will help identify biomass sources that achieve the greatest mitigation benefits.

To my knowledge, Canada has not deployed an independent national market modeling system in this context. Thus, the only viable option is to take a direct inventory-based approach (Smith et al. 2018). While the authors acknowledged that their study did not consider costs and other economic factors explicitly, they built alternative scenarios to capture the likely changes of environmental and economic conditions. It turns out that this kind of work is also data intensive. But given the data availability, this approach, coupled with reasonable assumptions, has worked well for Canada. Furthermore, the CFS researchers have closely followed the relevant principles of the Paris Agreement and the rules of the IPCC: their concentration on the managed forests suggests that they focused on anthropogenic ER&R and excluded changes induced by unmanaged natural forces. Also, their study clearly spelled out the baseline, additionality, and permanence. In a complementary study, Lemprière et al. (2017) reported that intensive forest management could account for 9.8% to 14.7% of Canada's annual $CO_2$-emissions reduction target of 112 million tons between 2014 and 2020, at a cost of less than $50 per ton.

*Europe*

Janssens et al. (2003) observed that most inverse atmospheric models reported considerable uptake of carbon dioxide in Europe's terrestrial biosphere. In contrast, carbon stocks in terrestrial ecosystems increased at a much smaller rate, with carbon gains in forests and grassland soils almost being offset by carbon losses from cropland and peat soils. Accounting for non-carbon dioxide carbon transfers that would not be detected by the atmospheric models and for $CO_2$ fluxes bypassing the ecosystem carbon stocks could reduce the gap between the small carbon-stock changes and the larger carbon dioxide uptake estimated by atmospheric models. The remaining difference could be because of missing components in the stock-change approach, as well as the large uncertainty in both methods.

With the use of the corrected atmosphere- and land-based estimates as constraints, the team found a net carbon sink between 135 and 205 teragrams (Tg) per year in Europe's terrestrial biosphere, which would be equivalent to 7 to 12% of the 1995 anthropogenic carbon emissions [~1.87 billion tons (Pg) of carbon]. Evidence also showed that the annual carbon sequestration of 0.72Pg in the European region corresponded to 9% of the annual European human-made carbon emissions. There would thus be a substantial potential to widen the accounting for biospheric uptake. So, Janssens et al. (2003) effectively identified a recovery of forest area and growing stock since the 1950s, following centuries of stock decline and deforestation, in Europe. Indeed, these re-growing forests were shown to be a persistent carbon sink, projected to continue for decades[3].

Just a decade later, alas, Nabuurs et al. (2013) reported that there were early signs of carbon saturation of European forests, claiming their detection of three warnings of saturation of the European forest biomass carbon sink. First, the volume increment rate is decreasing, and thus the sink is curbing after decades of increase. Second, land use is intensifying, thereby leading to deforestation and associated carbon losses. Third, natural disturbances are increasing and, consequently, so are the emissions of $CO_2$. Nabuurs et al. (2013) further emphasized that in managed European forests, carbon sink saturation seemed quite imminent. This would not be a sign of forest decline but rather an indication that these forests were reaching a dynamic equilibrium with the current intensity of management, tree species, and age-class distribution. Forest policies and strategies would thus need revision if the sink could be sustained. The onset of a saturation could be influenced by management in many ways: for example, by tree species, rotation length, age-class distribution, amount/type of harvesting, and so on. A sustainable management strategy aimed at maintaining or increasing forest carbon stocks, while producing an annual sustained yield of timber, fiber, or energy from the forest, would generate the largest mitigation benefit, but this should be applied in different ways throughout Europe, depending on the state of the forest. Further, countries should be less focused on the forest biomass sink strength and consider a mitigation strategy to maximize the sum of all the possible components: carbon sequestration in forest biomass, soil and wood products, and the effects of energy and material substitution of woody biomass.

*Russia*

Following the collapse of the USSR and the declining use of the Soviet Forest Inventory and Planning (FIP) system, Russia reported small changes in forested area, growing stock volume (GSV), and biomass to the UNFCCC and the FAO. According to the State Forest Register (SFR)—a primary repository of forest information and reporting apparatus to international agencies, the GSV and above ground biomass (AGB) increased by 1.1% and 0.6%, respectively, during 1990–2015. Yet studies using remote sensing indicated significantly increased vegetation productivity, tree cover, AGB, and forest ecosystem carbon. Schepaschenko et al. (2021) proclaimed that this inconsistency in estimates could be explained by an information gap that appeared when Russia decided to move from the FIP to another system for the collection of forest information at the national scale—the National Forest Inventory (NFI). According to the FIP report, the total GSV of Russian forests was $81.7 \times 10^9 m^3$ (without shrubland and with bias corrected). In contrast, the NFI, initiated in 2007, is a state-of-the-art inventory system based on a statistical sampling method. With its first cycle completed in 2020, the NFI data suggested that Russian forests accumulated $102 \times 10^9 m^3$ over its lifespan until 2014.

Schepaschenko et al. (2021) set out to provide an alternative estimate of the GSV and to assess the forest condition changes in post-Soviet Russia using two sources of information—ground measurements generated by the NFI, which are more accurate but have a limited number of observations, and remote sensing data products, which have a spatially comprehensive coverage but need careful calibration. They found the 2014 total GSV for the official forested area ($713.1 \times 10^6 ha$) to be $111 \pm 1.3 \times 10^9 m^3$, which is 39% higher than the $79.9 \times 10^9 m^3$ figure reported in the SFR. An additional $7.1 \times 10^9 m^3$ were found due to the enlarged forested area ($45.7\ 10^6 ha$), following the expansion of forests to the north, to higher elevations, in abandoned arable land, as well as the inclusion of parks, gardens, and other trees outside of forest, which were not counted as forest in the SFR. Assuming an unchanged total forest area (721.7 × 10$^6$ha) from 1988 to 2014, the authors showed that Russian forests had accumulated 1,163 × 10$^6$m$^3$ yr$^{-1}$ or 407Tg C yr$^{-1}$ in live biomass of trees over the 26 years[4]. This yields an average annual GSV change of 1.61m$^3$ ha$^{-1}$ or 0.56t C ha$^{-1}$.

To provide some context for the magnitude of these figures, they compared the Russian forest gain to the net GSV losses in tropical forests over the period 1990–2015. According to the Forest Resource Assessment reported by FAO (2015), an annual loss of 1,033 × 10$^6$m$^3$ yr$^{-1}$ in South and Central America, South and Southeast Asia, and Africa. In other words, Russia's gain in net forest stocking volume could more than offset the loss in the tropics. By multiplying the GSV density by the managed forest area (72%), Schepaschenko et al. (2021) further discovered that the GSV of managed forests was 94.2 × 10$^9$m$^3$, giving rise to a sequestration of 354Tg C yr$^{-1}$ in live biomass over 1988–2014, or 47% higher than reported in the national greenhouse gases inventory. The difference between their GSV estimate and that from the SFR report would be 23.6 × 10$^9$m$^3$. This study highlights that Russian forests play an even more important role in global carbon sequestration than previously thought.

## Discussion and summary

There exist alternative methodologies to carbon accounting and assessment in the forest sector, based directly or indirectly on forest inventory data, and they have varied strengths as they are applied under different circumstances. This chapter was aimed to review and synthesize the literature employing these approaches at the national and international levels. By linking up with a timber (or wood products) market model, an indirect inventory-based approach is predicated on economic theories of producers and consumers. As such, it can be more properly used to investigate the market responses (of producers and/or consumers) to changes in price, cost, and other economic incentives. On the other hand, a direct inventory-based approach can be readily deployed to project future forest conditions and their changes based on observed or assumed rates of growth, removal, mortality, and the like.

Regardless of economic or ecological responses, the primary variables of our interest include merchantable, stocking, and biomass volume, and carbon flux and stock. In addition to highlighting the potential uses and merits of different approaches, the case studies illuminate that deployed properly, either of them can generate powerful policy insights for forest management. Space limit does not permit me to include more cases of not only using different approaches but also reflecting different levels of aggregation in this chapter. In the next chapter, I will present a micro-level study of carbon accounting based on the pine plantation forests in the US South; and I will assess the national potential of China's forest sector in its decarbonization in Chapter 5.

As noted by Ohrel (2019), future forest GHG projection outcomes depend mainly on input data; model type and its specifications, scope, and spatial and sectoral coverage, and assumptions; and analytical objectives. Built for varying purposes, with different spatial and temporal scopes, capabilities, and different perspectives concerning future potential forest economic and environmental conditions, these approaches and models can sometimes produce divergent estimates of mitigation potential. These variations and divergences may make it difficult to discern consistent forest carbon dynamics and increase the complexity of establishing emissions reduction targets relative to historic values and tracking progress toward or deviation from those targets.

Within the direct inventory-based approach, different methods have their own strengths and weaknesses, too (Luyssaert et al. 2010). Current large-scale aggregate models or biogeochemical methods are not able to take local circumstances and/or forest management into account (Nabuurs et al. 2010). Having examined two types of models within the indirect inventory-based approach, Sjølie et al. (2015) reported that both were rather sensitive to changes in assumptions and data, and their strengths and weaknesses as well as appropriateness for responding to the research questions should be considered when choosing a modeling strategy. The authors went on to advise that if possible, using both models in parallel could remove the uncertainty caused by the foresight assumption and/or optimization routine and thus provides more useful information. As a matter of fact, both direct and indirect inventory-based approaches are needed, given their different orientations and strengths; whenever possible, they should complement each other in achieving the tasks of accounting and assessment.

In short, while highly aggregate, market-based modeling efforts remain indispensable to carbon accounting and assessment, however, more direct inventory-based, jurisdiction-oriented attempts are also needed. It is crucial for the forest sector modeling community to continually strive to update models with the most recent information and validate modeling outcomes by evaluating models and their outcomes independently and as part of larger model comparison efforts.

## Notes

1 See a definition of REDD+ in the Introduction.
2 See Chapter 1 for a discussion.
3 A similar finding was reported for some other regions of the Northern Hemisphere as well (Goodale et al. 2002).
4 They noted that the sequestration rate obtained should be treated with caution because different methods were applied in 1988 and 2014.

# References

Adams, D.M., Haynes, R.W. 1996. *The 1993 Timber Assessment Market Model: Structure, Projections, and Policy Simulations.* Gen. Tech. Rep. PNW-GTR-368. Portland, OR: U.S. Department of Agriculture, Forest Service, Pacific Northwest Research Station.

Austin, K.G., Baker, J.S., Sohngen B., Wade, C.M., Daigneault, A., Ohrel, S., Ragnauth, S., Bean, A. 2020. The economic costs of planting, preserving, and managing the world's forests to mitigate climate change. *Nature Communications* 11 (1): 1–9.

Baker, J., Sohngen, B., Ohrel, S., Fawcett, A. 2017. *Economic Analysis of Greenhouse Gas Mitigation Potential in the US Forest Sector.* RTI Press Publication No. PB-11–1708. Research Triangle Park, NC: RTI Press.

Buongiorno, J., Zhu, S., Zhang, D., Turner, J., Tomberlin, D. 2003. *The Global Forest Products Model: Structure, Estimation, and Applications.* Elsevier.

Daigneault, A., Johnston, C., Korosuo, A., Baker, J., Forsell, N., Prestemon, J.P., Abt, R.C. 2019. Developing detailed shared socioeconomic pathway (SSP) narratives for the global forest sector. *Journal of Forest Economics* 34 (1-2): 7–45.

Favero, A., Daigneault, A., Sohngen, B. 2020. Forests: Carbon sequestration, biomass energy, or both? *Scientific Advances* 6: eaay6792l.

Favero. et al. 2021. Assessing the long-term interactions of climate change and timber markets on forest land and carbon storage. *Environmental Research Letters* 16: 014051.

Food and Agriculture Organization of the United Nations (FAO). 2015. *Global Forest Resource Assessment Report.* Rome, Italy.

Food and Agriculture Organization of the United Nations (FAO). 2020. *Global Forest Resource Assessment Report.* Rome, Italy.

Goodale, C.L., Apps, M.J., Birdsey, R.A., Field, C.B., Heath, L.S., Houghton, R.A., Jenkins, J.C., Kohlmaier, G.H., Kurz, W., Liu, S.R., Nabuurs, G.J., Nilsson, S., Shvidenko, A.Z. 2002. Forest carbon sinks in the northern hemisphere. *Ecological Applications* 12 (3): 891–899.

Grassi, G., Stehfest, E., Rogelj, J., van Vuuren, D., Cescatti, A. et al. 2021. Critical adjustment of land mitigation pathways for assessing countries' climate progress. *Nature Climate Change* 11: 425–434.

Griscom, B.W., Adams J., Ellis P.W., Houghton R.A., Lomax G., Miteva D.A., Schlesinger W.H., Shoch, D. et al. 2017. Natural climate solutions. *PNAS* 114 (44): 11645–11650.

Guo, J., Gong, P., Brännlund, R. 2019. Impacts of increasing bioenergy production on timber harvest and carbon emissions. *Journal of Forest Economics* 34 (3-4): 311–335.

Hou, J.Y., Yin, R.S. 2022. How significant a role can China's forest sector play in decarbonizing its economy? *Climate Policy.* doi:10.1080/14693062.2022.2098229.

Houghton, R.A. 2008. Carbon flux to the atmosphere from land-use changes: 1850–2005. In: *TRENDS: A Compendium of Data on Global Change.* Oak Ridge, Tenn., U.S.A: Carbon Dioxide Information Analysis Center, Oak Ridge National Laboratory, U.S. Department of Energy.

Houghton, R.A., House, J.I., Pongratz, J., Van Der Werf, G.R., DeFries, R.S., Hansen, R.S., Quéré, C.L., Ramankutty, N. 2012. Carbon emissions from land use and land-cover change. *Biogeosciences* 9 (12): 5125–5142.

International Panel on Climate Change (IPCC). 2006. *Guidelines for National Greenhouse Gas Inventories.* Available at: www.ipcc.ch/report/2006-ipcc-guidelines-for-national-greenhouse-gas-inventories/.

International Panel on Climate Change (IPCC). 2019. *2019 Refinement to the 2006 IPCC Guidelines for National Greenhouse Gas Inventories*. Available at: www.ipcc.ch/report/2019-refinement-to-the-2006-ipcc-guidelines-for-nationalgreenhouse-gas-inventories/.

International Panel on Climate Change (IPCC). 2020. *Climate Change and Land: Summary for Policymakers*. Available at: www.ipcc.ch/srccl/.

International Panel on Climate Change (IPCC). 2022. *The Sixth Assessment Report*. Available at: www.ipcc.ch/assessment-report/ar6/.

Janssens, A., Freibauer, A., Ciais, P., Smith, P., Nabuurs, G.J., Folberth, G., Schlamadinger, B., Hutjes, R.W.A., Ceulemans, R., Schulze, D., Valentini, R., Dolman, D. 2003. Europe's terrestrial biosphere absorbs 7 to 12% of European anthropogenic $CO_2$ emissions. *Science* 300 (5625): 1538–1542. doi:10.1126/science.1083592.

Johnston, C., Buongiorno, J., Nepal, P., Prestemon, J. 2019. From source to sink: Past changes and model projections of carbon sequestration in the global forest sector. *Journal of Forest Economics* 34 (1-2): 47–72.

Johnston, C., Radeloff, V.C. 2019. Global mitigation potential of carbon stored in harvested wood products. *PNAS* 116: 14526–14531.

Jones, J.P.H., Baker, J., Austin, K., Latta, G., Wade, C., Cai, Y., Aramayo-Lipa, L., Beach, R., Ohrel, S., Ragnauth, S., Creason, J., Cole, J. 2019. Importance of cross-sector interactions when projecting forest carbon across alternative socioeconomic futures. *Journal of Forest Economics* 34 (3-4): 205–231.

Kaarakka, L., Cornett, M., Domke, G., Ontl, T., Dee, LE. 2021. Improved forest management as a natural climate solution: A review. *Ecological Solutions and Evidence* 2021 (2): e12090. https://doi.org/10.1002/2688-8319.12090.

Lan, J., Yin, R.S. 2017. Policy impact evaluation: Future contributions from economics. *Forest Policy and Economics* 83: 142–145.

Lemprière, T.C., Kurz, W.A., Hogg, E.H., Schmoll, C., Rampley, G.J., Yemshanov, D., McKenney, D.W., Gilsenan, R., Beatch, A., Blain, D., Bhatti, J.S., Krcmar, E. 2013. Canadian boreal forests and climate change mitigation. *Environmental Reviews* 21: 293–321.

Luyssaert, S., Ciais, P., Piao, S.L. et al. 2010. The European carbon balance. Part 3. Forests. *Global Change Biology* 16: 1429–1450. doi:10.1111/j.1365-2486.2009.02056.x.

Maziarz, M. 2022. Is meta-analysis of RCTs assessing the efficacy of interventions a reliable source of evidence for therapeutic decisions? *Studies in History and Philosophy of Science* 91: 159–167.

Mendelsohn, R., Sohngen, B.L. 2019. The net carbon emissions from historic land use and land use change. *Journal of Forest Economics* 34 (3-4): 263–283.

Millennium Ecosystem Assessment (MA). 2005. *Ecosystems and Human Well-being: Synthesis*. Washington, DC: Island Press.

Nabuurs, G.J., Hengeveld, G.M., van der Werf, D.C., Heidema, A.H. 2010. European forest carbon balance assessed with inventory based methods—An introduction to a special section. *Forest Ecol. Manage.* doi:10.1016/j.foreco.2009.11.024.

Nabuurs, G.J., Lindner, M., Verkerk, P.J., Gunia, K., Deda, P., Michalak, R., Grassi, G. 2013. First signs of carbon sink saturation in European forest biomass. *Nature Climate Change*. doi:10.1038/NCLIMATE1853.

Nordhaus, W. 2021. *The Spirit of Green*. New Jersey: Princeton University Press.

Ohrel, S. 2019. Policy perspective on the role of forest sector modeling. *Journal of Forest Economics* 34: 187–204.

Pagiola S., von Ritter, K., Bishop, J. 2004. *How Much Is an Ecosystem Worth?—Assessing the Economic Value of Conservation*. Washington, DC: The World Bank.

Piao, S.L., Fang, J.Y., Ciais, P., Peylin, P., Huang, Y., Sitch, S., Wang, T.et al. 2009. The carbon balance of terrestrial ecosystems in China. *Nature* 458, 1009–1013.

Piao, S.L., He, Y., Wang, X.H., Chen, F.H. 2022. Estimation of China's terrestrial ecosystem carbon sink: Methods, progress, and prospects. *Scientific China* (Earth Science) 65 (4): 641–651.

Rennert, K., Prest, B.C., Pizer, W.A., Newell, R.G., Anthoff, D., Kingdon, C., Rennels, L., Cooke, R., Raftery, A.F., Ševčíková, H., Errickson, F. 2021. *The Social Cost of Carbon: Advances in Long-Term Probabilistic Projections of Population, GDP, Emissions, and Discount Rates* (Brookings Papers on Economic Activity).

Roe, S., Streck, C., Obersteiner, M., Frank, S., Griscom, B., Drouet, L., et al. 2019. Contribution of the land sector to a 1.5 °C world. *Nature Climate Change* 9: 817–828. doi:10.1038/s41558-019-0591-9.

Schepaschenko, D., Moltchanova, E., Fedorov, S., Karminov, V., Ontikov, P., Santoro, M., See, L., Kositsyn, V., Shvidenko, A., Romanovskaya, A., Korotkov, V., Lesiv, M., Bartalev, S., Fritz, S., Shchepashchenko, M., Kraxner, F. 2021. Russian forest sequesters substantially more carbon than previously reported. *Scientific Reports* 11: 12825. https://doi. org/10. 1038/ s41598-41021-92152-92159.

Sjølie, H.K., Latta, G.S., Trømborg, E., Bolkesjø, T.F., Solberg, B. 2015. An assessment of forest sector modeling approaches: Conceptual differences and quantitative comparison. *Scandinavian Journal of Forest Research.* doi:10.1080/02827581.2014.999822.

Smith, C.E., Smiley, B.P., Magnan, M., Birdsey, R., Dugan, A.J., Olguin, M., Mascorro, V.S., Kurz, W.A. 2018. Climate change mitigation in Canada's forest sector: a spatially explicit case study for two regions. *Carbon Balance Management* 13: 11. https://doi.org/10.1186/s13021-018-0099-z.

Smith, C.E., Stinson, G., Neilson, E., Lemprière, T.C., Hafer M., Rampley, G.J., Kurz, W.A. 2014. Quantifying the biophysical climate change mitigation potential of Canada's forest sector. *Biogeosciences* 11, 3515–3529. doi:10.5194/bg-11-3515-2014.

Smith, P., Bustamante, M., Ahammad, H., Clark, H., Dong, H., Elsiddig, E.A. et al. 2014. Agriculture, forestry and other land use (AFOLU). In: Edenhofer, O., Pichs-Madruga, R., Sokona, Y., Farahani, Y., Kadner, S. et al. (eds) *Climate Change 2014: Mitigation of Climate Change. Contribution of Working Group III to the Fifth Assessment Report of the Intergovernmental Panel on Climate Change.* Cambridge and New York: Cambridge University Press.

Sohngen, B.L., Mendelsohn, R. 2003. An optimal control model of forest carbon sequestration. *American Journal of Agricultural Economics* 85: 448–457. doi:10.1111/1467-8276.00133.

Sohngen, B.L., Mendelsohn, R. 2007. A sensitivity analysis of forest carbon sequestration. In: *Human-induced Climate Change: An Interdisciplinary Assessment* (227–237). Cambridge:Cambridge University Press.

Sohngen, B.L., Mendelsohn, R., Sedjo, R. 1999. Forest management, conservation, and global timber market. *American Journal of Agricultural Economics* 81: 1–13.

United Nations. 2022. *COP27: Delivering for people and the planet.* Available at: www. un.org/en/climatechange/cop27).

United Nations Framework Convention on Climate Change (UNFCCC) Secretariat. 2020. *Reference Manual for the Enhanced Transparency Framework under the Paris Agreement.* Bonn, Germany.

Van Kooten, G.C. 2018. The challenge of mitigating climate change through forestry activities: What are the rules of the game? *Ecological Economics* 146: 35–43.

Wang, Y.F., Li, L.Y., Yin, R.S. 2021. A primer on forest carbon policy and economics under the Paris Agreement. *Forest Policy and Economics* 132 (102595).

Wear, D.N., Coulston, J.W. 2019. Specifying forest sector models for forest carbon projections. *Journal of Forest Economics* 34: 73–97.

Zeng, W.S., Tomppo, E., Healey, S.P., Gadow, K.V. 2015. The National Forest Inventory in China: History, results, and international context. *Forest Ecosystems* 2 (1). https://doi.org/10.1186/s40663-015-0047-2.

# 4 Outcomes of different frameworks in forest carbon accounting at the local level

**Introduction**

It is generally accepted that if protected and managed adequately, forests can play a significant role in reducing the emission of $CO_2$ and removing it from the atmosphere and do so in a cost-effective manner (Seymour and Busch 2016, Fuss et al. 2018, Nordhaus 2021). Thus, nature-based solutions (NbSs) by the forest sector are widely recognized as an important pathway for mitigating climate change (IPCC 2020, Roe et al. 2019). But advancing along this pathway calls for major research efforts to be devoted to forest carbon accounting and assessment, which can be conducted at different levels of aggregation (UNFCCC 2020, IPCC 2006), as discussed in the last chapter. This chapter is motivated to conduct a carbon accounting and thus offset credit determination at the local or micro level.[1]

Unlike studies at national or even higher aggregate level (e.g., Baker et al. 2019, Austin et al. 2020), this kind of micro-level work is directly relevant to the participation of individual forestland owners in NbSs. Studying forest sector carbon sequestration and storage ($S_1$ & $S_2$) at the micro level has attracted attention from many scholars, including economic and policy analysts (e.g., van Kooten et al. 1995, Tahvonen and Kallio 2006, Ekholm 2016, van Kooten 2018). Substantive research progress notwithstanding, there remain deficiencies over how to incorporate $S_1$ & $S_2$ into carbon accounting and policy assessment properly (Wang et al. 2021).

As elaborated below, one of the deficiencies is reflected in the frequently used Hartman modification to the Faustmann model for determining the optimal rotation age and land expectation value of a timber stand (van Kooten et al. 1995, Tahvonen and Kallio 2006), which is not well-suited for dealing with a continuous process underlying forest carbon accounting or policy assessment (Li et al. 2020). Predicated on the adoption of such a discrete model, existing studies have largely failed to conform to the accounting principles of the Paris Agreement, treat timber and carbon as joint products appropriately, and capture the differentiated potentials of carbon storage by harvested wood products (HWPs).

DOI: 10.4324/9781003436652-5

Therefore, it is imperative to establish a more satisfactory analytic framework of forest carbon accounting and explore the outcomes of alternative frameworks and their likely deviations. To that end, my work is built on a forest-level profit function that treats timber and carbon as joint products coherently and captures the Paris Agreement accounting principles and the differentiated potentials of carbon storage in HWPs. It is hoped that in combination, these steps will contribute to an improved handling of forest sector NbSs and thus an enhanced understanding of their implications to forest management and wood utilization.

Using an intensively managed pine plantation in the US South, my empirical case demonstrates that the outcomes indeed deviate a lot, with those derived from the Hartman-Faustmann framework exaggerating the amount of carbon offset credits a landowner can take by a factor of at least 2.76. Also, while adding carbon to the valuation has limited impact on extending the optimal rotation age, its effect on raising the net revenue can be substantial. I believe that these findings are of broad interest, given the great desire for individual landowners to participate in forest sector climate actions in all parts of the world (FAO 2016, IPCC 2020).

The rest of this chapter is organized as follows. First, I examine the deficiencies of the literature as well as the concrete ways to move the research agenda forward. Then, I consider the conceptual issues revolving around forest carbon accounting and joint production. Next, using a pine plantation example of the US South, I show the implications of the requisite accounting principles when timber and carbon are treated as joint products. Finally, I summarize my analysis and discuss my findings.

## Literature review

In most local-level economic and financial studies, carbon sequestered by trees has been treated as another output of the joint forest production, which is a process that results in more than one output using a common input, or common inputs (Bowes and Krutilla 1989). When dealing with timber alone, the familiar Faustmann model—a discounted net benefit formulation for a stand of trees over an infinite time horizon—is often used to determine the optimal rotation age and land expectation value. Accordingly, in studying joint forest production, a modification to the Faustmann model is made by adding another term for the net non-timber benefit (carbon in this case). Since this kind of modification was originally introduced by Hartman (1976), it is often called the Hartman modification (van Kooten et al. 1995, Hyde 2012).

Compared to the timber-only scenario, incorporating both timber and carbon into the valuation affects a landowner's management outcome and decision-making. For instance, Ekholm (2016) shows that when starting from bare land, the initial carbon price and its growth rate both increase the length of the first rotation. Hou et al. (2020) claim that the optimal rotation age increases for all examined species when considering the joint production of timber and carbon. By distinguishing "temporary" Certified Emission Reductions (tCER covering 20 years) and "long-term" Certified Emission Reductions (lCER covering 30 years) under the Kyoto Protocol, the authors also find that accounting regimes can have a large impact on the net economic benefit.

Further, earlier studies have recognized the importance of such carbon pools as HWPs and dead organic matter (DOM) on the forest floor, in addition to standing trees. After presenting a conceptual framework for integrating multiple carbon pools, Asante and Armstrong (2012) demonstrate that incorporating DOM and HWP have the effect of reducing the rotation age. Perhaps more interestingly, when initial stocks of carbon in these pools are relatively high, considering them can have a highly negative effect on net present value. On the other hand, Holtsmark et al. (2013) discover that Asante and Armstrong (2012) overlooked carbon released from decomposition of DOM and HWPs following the harvest. When this has been corrected, the sizes of the initial stocks of DOM and HWPs do not influence the optimal rotation age. Moreover, that study suggests that including DOM leads to longer, not shorter, rotation periods.

Notably, Gutrich and Howarth (2007) and van Kooten (2018) are among the few studies that were able to consider the life cycle of carbon through the vertical chain of processing wood. The former finds that accounting for the value of carbon sequestration favors longer (in some cases much longer) optimal rotation periods and/or the adoption of partial harvesting regimes. The latter exhibits how to decide which forestry activities generate carbon offset credits, with a stress over the centrality of the "rules of the carbon game" to determining the number of offset credits; nevertheless, it did not explicitly consider the Paris Agreement's accounting requirements itself. As interesting and insightful as these studies may be, they closely followed the modeling strategy of van Kooten et al. (1995) in exploring the outcomes of alternative scenarios of forest management and wood products manufacturing, based on a Hartman modification to the Faustmann model.

Also, certain parameters used, including the wood decay rate and the portion of timber entering the product pool, are at odds with the IPCC rules. Indeed, when defining the scope of forest sector mitigation actions and enumerating their magnitudes, what van Kooten (2018) referred to was the dated IPCC 2000 rules instead of the more comprehensive *2006 Guidelines for National Greenhouse Gas Inventories*. Because of its much earlier publication, of course, it was impossible for Gutrich and Howarth (2007) to heed to the IPCC 2006 guidelines or the Paris Agreement carbon accounting principles, particularly additionality and permanence, as already noted in Chapters 1 and 2. In the end, their numerical examples look more like a regional-level assessment of forest sector potential of $S_1$ & $S_2$ rather than a local-level accounting that I am concerned about in this chapter.

In any event, a Hartman modification to the Faustmann model may not be a very effective framework for forest carbon accounting or policy assessment because of its intrinsic limitations. The most salient limitation is that as a stand-level formulation, the Faustmann model features a discrete, point-input and point-output production process that is not able to explicitly represent the forest-level, continuous production process (Li et al. 2020, Yin and Newman 1997). The latter is, however, more compatible with the requirements of carbon accounting and the jurisdictional governance of climate change mitigation (Wang et al. 2021).

More specifically, the Paris Agreement requires that carbon sequestered in forest ecosystems (FESs) and stored in HWPs be additional and permanent as measured against an established baseline (FAO 2016, Galik et al. 2016). The baseline emission or removal level is usually defined by a country in its NDC (nationally determined contributions) and applied to various entities under its jurisdiction (Wang et al. 2021). If additionality or permanence cannot be ensured, it is invalid to take full credit for emission offsets from estimated carbon stock increase. Consequently, it is inappropriate to assume financial rewards from the calculated gains of carbon stock and storage at the market price (in $CO_2$ equivalent) of traded emission allowances (van der Gaast et al. 2018). Thus, a proper starting point for carbon accounting and policy assessment is to work at the forest level, if not a more aggregate one (Li et al. 2020).

In addition, the concept of joint production in forestry itself has not been scrutinized (Irland and Sample 1995). Unlike the cases of timber vs. water or timber vs. non-timber forest product (NTFP), timber and carbon are both biomass-based and thus inseparable. As a result, once harvested, the fate of timber dictates that of carbon, given that the timber may be destined for sawnwood, wood-based panels, or paper/paperboard. Because the potential of harvested timber producing these outputs is primarily determined by the (mean) diameter of trees, which is in turn a function of their (mean) age, the proportion of different products varies with time, all else being equal. These factors will ultimately influence the quantity and life of carbon contained in HWPs (IPCC 2006).

In sum, it is essential to incorporate forest carbon $S_1$ & $S_2$ properly into a local accounting or assessment. For that purpose, it is needed to:

1 make the analytic framework more compatible with the basic accounting principles of the Paris Agreement;
2 treat timber and carbon appropriately as joint forest products; and
3 give sufficient attention to the varied potentials of carbon storage in different HWPs.

I believe that these steps pertain to the credibility of any relevant research attempt and its findings, and that the time has come for forest economics and policy analysts to build them into a harmonized, comprehensive analytic framework.

## Conceptual considerations

### Accounting requirements

A forest sector NbS can be accounted for only when it represents a reduction or removal of anthropogenic emissions relative to an established *baseline* (UNFCCC 2020). Therefore, a key first step is to define a Party's baseline, against which the *additionality* of its concrete actions in the forest sector can be determined. In a net–net accounting, total net emissions or removals in the base year are subtracted from total net emissions or removals in the accounting period to find the total amount of credits or debits resulting from these actions (Wang et al. 2021).

Any forest sector mitigation action may lead to non-*permanence* if it is not well designed. This is mainly because forest sector actions inevitably entail long-term land use, management commitment, and/or attempting to make the HWPs last as long as possible. As already stated, the duration of an action should be commensurate with the life of an NDC; certainly, it must be longer and more coherent than what was practiced under the Kyoto Protocol notions of tCER and lCER (FAO 2016). Given the current consensus for realizing global carbon neutrality by mid-century (Ou et al. 2021), forest sector $S_1$ & $S_2$ must last at least until then. In fact, the mainstream convention is to consider the effectiveness of any major action until the end of this century (Fuss et al. 2018, Qin et al. 2021). To ensure performance, the specific effects of capturing various carbon pools and their corresponding durations on the accounting and management outcomes must be carefully measured and monitored.

### Aggregate process

Any sensible attempt of forest carbon accounting or policy assessment must look beyond a single stand by examining a more aggregate, forest-level production process, which is possibly the smallest unit compatible with the jurisdictional governance approach (Wang et al. 2021). This is not because the stand-level Faustmann model is wrong per se; rather, it is because the forest-level, aggregate unit is more congruent with the nature of forest sector mitigation actions and the corresponding requirements for any NbS, namely additionality and permanence measured against a baseline.

At the stand level, harvesting and regeneration decisions are made intermittently, and the regeneration decision depends on the harvest decision, which is why the Faustmann model is considered a point-input and point-output model (Yin 1997). At the forest level, on the other hand, harvesting, regeneration, and management decisions must be made simultaneously and continuously to sustain timber production and/or carbon sequestration. Also, it is known that the stand-level Faustmann model does not allow an explicit consideration of simultaneous shifts at both the extensive and intensive margins of forest production (Yin 1997, Hyde 2012). As a result, Wang et al. (2021) called for "reorienting the inquiry into forest C[carbon] policy and economics from the predominantly disaggregated, project-based paradigm to a more aggregated, program-oriented one, such that it will be more congruent with and conducive to the NDC structure and the jurisdictional approach."

It should be reiterated that while a forest-level production process is perhaps the smallest unit of carbon accounting compatible with the NDC-centered architecture of the Paris Agreement, it is not the only one. Regional, national, and international assessments of the forest sector's potential for carbon emission reduction and removal have been carried out (e.g., Austin et al. 2020, Baker et al. 2019, Wear and Coulston 2019). But this study takes a local-level perspective, which I argue is more foundational not only for carbon accounting but also for offsetting and credit taking.

*Joint production*

In cases of joint production like timber vs. non-timber products or timber vs. water, the products, albeit closely related, are separate (Bowes and Krutilla 1989). That is, timber, water, and an NTFP can exist independently (Swallow et al. 1990). In contrast, in the case of timber vs. carbon, carbon is embodied in the timber and thus the product(s) derived from timber (IPCC 2006, van Kooten et al. 1995). However, different HWPs have different life spans, causing the carbon stored in them to be released at different rates—a critical consideration for the permanence condition of emission offsetting and credit taking.

IPCC (2006) has made it clear that timber harvesting is partitioned into "roundwood" (or "wood-removals") and "slash" (generally left in the forest). Roundwood is then further subdivided into "industrial roundwood" and fuelwood or charcoal. "Industrial roundwood" can in turn be used to manufacture a combination of the three wood product commodity classes: sawnwood, wood-based panels, and paper/paperboard. Accordingly, the default half-lives of the three semi-finished HWP classes are: 35 years for sawnwood, 25 years for wood-based panels, and two years for paper/paperboard. It is thus crucial to consider the life cycles of different HWPs when scrutinizing the content and duration of carbon embodied in them.

## An illustrative case

Here, I illustrate the conceptual advances I have articulated using a case of southern pine management in the US. First, I present my empirical model, a profit function of joint forest production. Next, I list the data sources and characterize the forest used for the case study. Following these steps, I report my findings.

*Analytic framework*

Let me begin by considering timber as the only output of forest production. Given that input and output decisions in the production process change in response to evolving market conditions, I can express the profit maximization as:

$$\pi_t(p_t, i_t, w_t, r_t) = max[P_t Q(t) - i_t P_t I(t) - w_t K(t) - l_t L(t)],$$

where a timber producer incurs an initial investment expense at $t = 0$ in establishing its forest inventory $I(0)$; thereafter, facing a market price $P_t$ for timber, a unit regeneration cost $w_t$, a land rental cost $l_t$, and a discount rate $i_t$, this produces a flow of outputs $Q(t)$ in the future as additional annual operating inputs $K(t)$ and a land acreage equal to $L(t)$ are committed, and an inventory level of $I(t)$ is maintained. This analytic framework captures the fundamental long-run, continuous nature of forest production, which can be viewed as a process that employs operating inputs and land, along with the existing inventory, to achieve a new level of inventory that is then used to produce a certain amount of stumpage (Li et al. 2020).

When both timber and carbon are treated as outputs, the above model must be modified. First, the total revenue is now a sum of timber revenue $P_{1t}Q_1(t)$ and carbon revenue $P_{2t}Q_2(t)$, where $P_{1t}$ and $P_{2t}$, and $Q_1(t)$ and $Q_2(t)$ are, the prices and quantities of the two classes of outputs, respectively. Furthermore, it is sensible to modify the opportunity cost associated with holding an inventory of $I(t)$, which I will do in my numerical scenarios. Note that this opportunity cost should be weighted by the prices of the two classes of outputs corresponding to their statuses and dynamics.

Given the need for urgent actions to alleviate climate change, an argument can be made to put a time value on carbon sequestered in FES and stored in HWPs (van Kooten 2018, Sohngen et al. 2022). But I feel that it is more complicated than that, because estimating the social cost of carbon (SCC) should consider the time value of emission reduction as well as removal (Rennert et al. 2021). That is, the SCC is supposed to capture the value of delayed emission or accelerated sequestration through properly discounting the accumulative monetary damages—earlier and more significant actions leading to a lower SCC. As shown below, also, the rotation age of a pine plantation is relatively short anyway; thus, the added time value may not have a sizable impact on landowners' decision making.

*Data sources*

The forest production process and its associated input use and output generation of my case study is based on a scenario explored by Yin et al. (1998)—the only dataset I possess. The advantage of the dataset is that it has all the relevant information on both the growth and yield under alternative regimes of silvicultural practices and the corresponding inputs used and outputs generated over time, as well as statistics for input costs and output prices. The growth-and-yield data originated from a long-term experimental study conducted by the University of Georgia Plantation Management Research Cooperative, installed in 1979 on cutover land after the existing slash pine plantations had been harvested. The Cooperative replicated silvicultural treatments at 16 locations along the lower coastal region of Georgia and

68  *Outcomes of different frameworks*

northern Florida (Pienaar and Rheney 1995). Each treatment plot covered 1/2 acre with the interior 1/5-acre plot being measured.

The average site index was estimated at 60 feet, and the planting density was uniformly 600 trees/acre. Each site was prepared with the standard treatment for machine planting (S), consisting of a single pass with a rolling-drum chopper after harvesting, followed by a broadcast burn before planting. Then, combinations of the following three treatments were adopted: bedding (B)—double-pass with bedding plow before planting; fertilization (F)—application of 250lb/acre of ammonium phosphate after the 1$^{st}$ growing season, followed with 150lb/acre of nitrogen, 50lb/acre of phosphorus, 100lb/acre of potassium broadcast after year 12; and herbicide use (H)—treatment to control competing vegetation after standard chopping and burning.

*Forest characterization*

Let me focus only on the "S+B+H" regime—standard machine planting plus bedding and herbicide use, with its growth process and productivity summarized in Table 4.1. The price and cost parameters are listed in Table 4.2. There exist local markets for both sawtimber and pulpwood, and these products are sold at different prices. Meanwhile, the growth/yield models used determine the amounts of these products for a stand of trees at a given point in time and thus show their changes over time. For analytic convenience, a regulated forest is also assumed in dealing with the flow and stock variables of carbon accounting; thus, stands of equal sizes but different ages are distributed over space.

Table 4.1 shows that compared to pulpwood, sawtimber appears later in the rotation and amounts to only 23%, 27%, and 44% of the total merchantable volume, respectively, at age 15, 20, and 25. Likewise, the amounts of merchantable timber and $CO_2$ sequestered per acre in a stand of trees at age 15, 20, and 25 are 2804.1ft$^3$ and 50.6 tons, 3794.0ft$^3$ and 68.4 tons, and 4360.4ft$^3$ and 78.7 tons, respectively. The post-harvest slashes are often burned as part of the site preparation before planting the next crop of trees, and the soil carbon and DOM pools are stable and vary little over time (Pienaar and Rheney 1995).

As reported in Table 4.3, the oldest stand reaches financial maturity for harvesting (clear-cut) when it is 20 years old if only timber production is valued, with sawtimber and pulpwood outputs of 804.2ft$^3$/acre and 2450.5ft$^3$/acre. Based on the price and cost parameters listed in Table 4.2 and an annual discount rate of 6%, the total revenue, total cost, and net revenue are $2039.7, $1634.8, and $404.9 per acre per year, respectively. Such a 20-year-old stand contains 68.38 tons/acre of $CO_2$ in its biomass of merchantable volume (40.34 tons in pulpwood and 28.04 tons in sawtimber), and the total stocking volume of the whole forest carries 521.22 tons of $CO_2$ in its standing trees of different ages. Once the 20-year-old stand is harvested, however, the amount of $CO_2$ contained in the HWPs on a permanent basis—12.14 tons in sawnwood and 1.52 tons in paper/paperboard—becomes much smaller. Because of the irrecoverable residuals generated in making

Table 4.1 Estimated growth and yield for the standard plus bedding and herbicide use regime (S+B+H) of slash pine plantation forests in the US South

| t | N | H | B | V | $V_1$ | $V\_st$ | $V\_pw$ | $V\_st/V_1$ | $V\_pw/V_1$ | $CO_2$ |
|---|---|---|---|---|---|---|---|---|---|---|
| 1  | 600.0 | 2.5  | 2.5   | 1.0    | 0.0    | 0.0    | 0.0    | 0.00 | 1.00 | 0.0  |
| 10 | 550.2 | 35.0 | 84.2  | 1528.7 | 1415.1 | 21.7   | 1393.5 | 0.02 | 0.98 | 25.5 |
| 15 | 515.2 | 49.9 | 117.1 | 2880.5 | 2804.1 | 639.6  | 2164.5 | 0.23 | 0.77 | 50.6 |
| 20 | 478.4 | 60.9 | 132.9 | 3854.9 | 3794.0 | 1568.7 | 2225.3 | 0.27 | 0.73 | 68.4 |
| 21 | 471.0 | 62.7 | 134.6 | 3997.5 | 3938.8 | 1731.9 | 2206.9 | 0.31 | 0.69 | 71.1 |
| 22 | 463.6 | 64.4 | 136.0 | 4123.7 | 4067.0 | 1883.4 | 2183.6 | 0.35 | 0.65 | 73.4 |
| 23 | 456.2 | 65.9 | 137.0 | 4234.2 | 4179.4 | 2023.0 | 2156.4 | 0.38 | 0.62 | 75.4 |
| 24 | 448.8 | 67.3 | 137.7 | 4330.0 | 4276.9 | 2150.6 | 2126.3 | 0.41 | 0.59 | 77.2 |
| 25 | 441.4 | 68.6 | 138.2 | 4411.8 | 4360.4 | 2266.4 | 2093.9 | 0.44 | 0.56 | 78.7 |
| 26 | 434.0 | 69.8 | 138.4 | 4480.8 | 4430.9 | 2371.0 | 2059.8 | 0.46 | 0.54 | 79.9 |
| 27 | 426.6 | 70.9 | 138.3 | 4537.7 | 4489.3 | 2464.9 | 2024.3 | 0.48 | 0.52 | 81.0 |
| 28 | 419.3 | 72.0 | 138.1 | 4583.5 | 4536.5 | 2548.7 | 1987.8 | 0.50 | 0.50 | 81.8 |
| 29 | 412.0 | 72.9 | 137.8 | 4619.2 | 4573.5 | 2623.1 | 1950.5 | 0.52 | 0.48 | 82.5 |
| 30 | 404.8 | 73.8 | 137.3 | 4645.6 | 4601.2 | 2688.7 | 1912.5 | 0.54 | 0.46 | 83.0 |

Notes:
1) The data were derived from Yin et al. (1998), and "S+B+H" represents the regime of standard machine planting plus bedding and herbicide use.
2) t = year; N = number of trees per acre; H = dominant height (feet); B = basal area (ft²/acre); V = total volume (ft³/acre); $V_1$ = total merchantable volume (ft³/acre); $V\_st$ = sawtimber volume (ft³/acre); $V\_pw$ = pulpwood volume (ft³/acre); and $CO_2$ = amount of carbon dioxide sequestered (ton/acre).
3) In estimating the amount of carbon sequestered, I used the following conversions: merchantable volume × 1.12 = biomass volume (softwood); 1 cubic foot = 0.0078125 cord; 1 cord biomass volume = 5350 pounds = 2.675 green tons; biomass green ton tons = 0.463 dry tons; 1 metric ton = 1.10231 short tons; dry biomass ton = 0.5 carbon ton; and 1 ton of carbon = 3.67 tons of carbon dioxide.

70  Outcomes of different frameworks

Table 4.2 Input cost and output price data of forest production in the US South

| Sawtimber price | $113.80/cord | Bedding | $48/acre |
|---|---|---|---|
| Pulpwood price | $37.10/cord | Fertilization | $150/acre |
| Land rental cost | $350/acre | Herbicide | $35/acre |
| Standard treatment | $88/acre | Overhead | $2/acre/year |
| Mechanical planting | $55/acre | Ad valorem tax | $2.5/acre/year |
| Seedlings | $28/acre | $CO_2$ price | $20/ton |

*Notes:*
$CO_2$ price is based on the recent emission trading information in California (www.bloomberg.com/news/articles/2021-11-24/california-carbon-sets-record-price-in-cap-and-trade-auction). Other price and cost data are from Yin et al. (1998).

the products and the extremely short duration of paper/paperboard, this leads to over 80% of carbon sequestered in sawtimber and pulpwood quickly emitted back to the atmosphere.

## Main empirical results

Depending on the assumptions used, numerous management scenarios can be examined. Here, I have decided to concentrate on a few more representative scenarios in terms of whether and how the additionality and permanence principles are followed, in conjunction with the status of plantation establishment. First, I assume that for the country within which the landowner (or group of owners) resides, the baseline of its NDC is year 2000, and 2050 is the end of its commitment. The additionality principle implies that the stocking volume of any of the forest stands that have been established by 2000 are not eligible for emission offsetting and/or credit taking. As a result, it becomes necessary to consider whether the plantation is fully, partially, or not yet established by the base year. Likewise, the permanence principle dictates that once a landowner decides to join a carbon emission offsetting initiative, s/he will have to stay in it until 2050.

The annual timber removals consist of sawtimber used for producing sawnwood, and pulpwood used for making paper/paperboard. Again, the IPCC guidelines indicate that sawnwood has an average service half-life of 35 years, whereas paper/paperboard has an average service half-life of only two years. Further, sawtimber has an incremental rate of conversion into sawnwood, ranging from 55% at a stand age of 15 years to 65% at a stand age of 25 years or more. The residuals generated from converting sawtimber to sawnwood is used for producing pulp, which is then turned into paper/paperboard, with a recovery rate of 95%.

When both carbon and timber are included in valuation with the given price, cost, and discount rate parameters, the optimal rotation age remains at 20 years, as shown in Table 4.3. If the carbon price goes up from $20 to $50 per ton of $CO_2$, the rotation age increases by only one year. Thus, considering

Outcomes of different frameworks 71

Table 4.3 Estimated annual revenues, costs, profit, and timber and carbon outputs per acre for the standard plus bedding and herbicide use regime (S+B+H)

| t | Revenue PW | Revenue ST | Revenue Total | Cost Cap | Cost Land | Cost Oper | Cost Total | Net Revenue | Output V_pw | Output C_pw | Output V_st | Output C_st |
|---|---|---|---|---|---|---|---|---|---|---|---|---|
| \multicolumn{13}{l}{Timber only} |
| 19 | 648.42 | 1240.10 | 1888.53 | 722.95 | 470.00 | 294.00 | 1486.95 | 401.57 | 17.5 | | 10.90 | |
| 20 | 644.99 | 1394.71 | 2039.70 | 845.34 | 493.50 | 296.00 | 1634.84 | 404.86 | 17.4 | | 12.26 | |
| 21 | 639.66 | 1539.76 | 2179.43 | 976.10 | 517.00 | 298.00 | 1791.10 | 388.32 | 17.2 | | 13.53 | |
| \multicolumn{13}{l}{Timber and Carbon (I: carbon price is \$20 per ton of $CO_2$)} |
| 19 | 679.07 | 1452.70 | 2131.77 | 796.09 | 470.00 | 294.00 | 1560.09 | 571.67 | 17.5 | 1.53 | 10.90 | 10.63 |
| 20 | 675.47 | 1637.53 | 2313.01 | 934.87 | 493.50 | 296.00 | 1724.37 | 588.63 | 17.4 | 1.52 | 12.26 | 12.14 |
| 21 | 669.89 | 1811.97 | 2481.86 | 1083.78 | 517.00 | 298.00 | 1898.78 | 583.08 | 17.2 | 1.51 | 13.53 | 13.61 |
| \multicolumn{13}{l}{Timber and Carbon (II: carbon price is \$50 per ton of $CO_2$)} |
| 20 | 721.20 | 2001.77 | 2722.97 | 1069.18 | 493.50 | 296.00 | 1858.68 | 864.29 | 17.4 | 1.52 | 12.26 | 12.14 |
| 21 | 715.24 | 2220.27 | 2935.51 | 1245.31 | 517.00 | 298.00 | 2060.31 | 875.20 | 17.2 | 1.51 | 13.53 | 13.61 |
| 22 | 707.67 | 2425.75 | 3133.42 | 1433.31 | 540.50 | 300.00 | 2273.81 | 859.60 | 17.1 | 1.50 | 14.71 | 15.03 |

Notes:
1) "S+B+H" represents the regime of standard machine planting plus bedding and herbicide use.
2) $t$ = year; $PW$ = pulpwood revenue; $ST$ = sawtimber revenue; $Cap$ = capital; and $Oper$ = operating input cost. Revenue and cost terms are in US\$/acre. An annual discount rate of 6% was used in the estimation.
3) $V\_pw$ = pulpwood volume (cord/acre); $V\_sw$ = sawtimber volume (cord/acre); $C\_pw$ = amount of carbon dioxide sequestered in pulpwood (ton/acre); and $C\_st$ = amount of carbon dioxide sequestered in sawtimber (ton/acre). $C\_pw$ and $C\_st$ were derived from harvested wood products on a permanent basis—lasting until 2050.
4) The rows in red correspond to the rotation age when the net revenue is maximized.

carbon sequestration by participating in an offsetting initiative has limited effect on the rotation age. This is by and large because if carbon stored in HWPs is calculated on a permanent basis and valued at a relatively low price, the same economic calculus—postponing timber harvesting would mean the incurrence of a higher opportunity cost—still holds. However, including carbon value has a more pronounced effect on profitability. With a price of $20 per ton of $CO_2$, the net revenue rises from $404.9 to $588.6 per acre per year—a more than 45% gain compared with the timber-only scenario. Likewise, when the price goes up $50 per ton of $CO_2$, the net revenue increases to $875.20 per acre per year.

To save space, I decided not to further explore the sensitivity of my findings to the changed discount rate, prices of sawtimber and pulpwood, or costs of inputs in this chapter. A curious reader can pursue this task based on the data file I have provided in the Supplementary Information of Sun et al. (2023). The following specific scenarios, shown in Table 4.4, correspond to the three different stages of the plantation establishment and the costs, prices, and discount rate listed in Table 4.2.

In sum, my results show that the outcomes of forest carbon accounting are determined by whether the additionality and permanence principles are followed, where the plantation establishment stands, and how the HWPs are treated, in addition to the basic economic factors—prices, costs, and discount rate. The estimated carbon credits vary dramatically under different combinations of these considerations.

## Discussion and summary

This chapter was intended to overcome some of the analytic challenges encountered in the current deliberation and practice of forest carbon accounting and credit taking at the micro level. These challenges center on whether an accounting exercise follows the Paris Agreement's additionality and permanence principles for carbon offsetting, how timber and carbon are incorporated into the process of forest production, and if the accounting captures varied magnitudes and durations of HWPs. Inadequate attention or an inappropriate framework to address these issues can cast doubt on the validity of any effort in promoting forest sector NbSs.

My analysis has demonstrated that properly accounted for, forest carbon offsetting actions are important and credit worthy and thus deserve to be seriously promoted and financially rewarded. Because of the embodiment of carbon in timber, as well as the relatively low carbon prices, however, "it is not necessarily a choice of either timber or C (carbon) in reality; rather, it is a matter of both in many cases" (Li et al. 2020). But incorporating carbon into the valuation can alter the logic and outcome of forest management. Determining the rotation length must now pay proper attention to HWPs because of their markedly different service lives and the disproportionality between carbon stored in sawnwood and that in paper/paperboard. It is thus vital to

*Outcomes of different frameworks* 73

*Table 4.4* Estimated carbon sequestration/storage outcomes (tons) under different scenarios for the S+B+H regime (carbon price is $20 per ton of $CO_2$)

| Scenario | $CO_2$ in standing trees | $CO_2$ in annual HWPs | | Offsetting credits in 50 years | |
|---|---|---|---|---|---|
| | | Pulpwood | Sawntimber | Claimed | Acceptable |
| 1 | **If the additionality and permanence principles are followed** | | | | |
| A | Plantation already fully established | | | | |
| | 0.00 | 1.52 | 12.14 | 683.00 | 683.00 |
| B | Plantation needs 10 more years to be fully established | | | | |
| | 471.43 | 1.52 | 12.14 | 1,017.83 | 1,017.83 |
| C | Plantation begins to be established after the base year | | | | |
| | 521.22 | 1.52 | 12.14 | 931.02 | 931.02 |
| 2 | **If the additionality and permanence principles are not followed** | | | | |
| A | Plantation already fully established | | | | |
| | 521.22 | 40.34 | 28.04 | 3,930.22 | 683.00 (same as 1-A) |
| B | Plantation needs 10 more years to be fully established | | | | |
| | 521.22 | 40.34 | 28.04 | 3,256.42 | 1,017.83 (same as 1-B) |
| C | Plantation begins to be established after the base year | | | | |
| | 521.22 | 40.34 | 28.04 | 2,572.60 | 931.02 (same as 1-C) |

*Notes:*
A. "S+B+H" represents the regime of standard machine planting plus bedding and herbicide use.
B. The amount of $CO_2$ sequestered in standing trees is an outcome of the total stocking volume, which is a sum of the volumes of stands in different ages, multiplied by the relevant conversion factors.
C. In calculating the amount of $CO_2$ stored in HWPs, I considered sawnwood produced from sawtimber, and paper/paperboard produced from pulpwood along with the residual material generated in sawnwood production.
D. The commitment period for a landowner is 50 years (2000–2050) to comply with the permanence requirement for taking carbon credits.
E. If the whole plantation was established before the base year, then it is not eligible for inclusion in the offsetting accounting, and the only eligible offsetting credits are derived from the annual removal of a mature stand, assuming the annual growth is removed entirely. Under these circumstances (scenario 1-A), the additional accountable amount of stored carbon would be 13.66 tons per acre per year (i.e., 1.52 tons in paper/paperboard and 12.14 tons in sawnwood), or 683.00 tons per acre over the 50 years. On the other hand, ignoring the additionality and permanence requirements, one could mistakenly assume a $CO_2$ offset of as much as 3,930.2 tons, including 521.22 tons in standing trees and 68.38 tons a year in pulpwood and sawtimber produced over 50 years (scenario 2-A).
F. If the plantation has not been fully established by the base year, the landowner must complete the plantation development first. If the plantation establishment began 10 years ago, the landowner needs to take another 10 years to get it fully established before starting timber harvesting. In this case, part of the carbon stock of the plantation gained after the base year (521.22–49.82 = 471.40 tons) and the annual removals (13.66 tons) over the remaining 40 years (546.4 tons) until 2050 can be counted as offsetting credits. By the end of the commitment period, the accountable amount of $CO_2$ would total 1017.8 tons (scenario 1-B). If the additionality and permanence principles were disregarded, the landowner might erroneously claim credits from an accumulated $CO_2$ offset of 3,256.4 tons (scenario 2-B)—521.22 tons in standing trees and 68.38 tons a year from pulpwood and sawtimber produced over the 40 years.
G. If the plantation establishment has not yet begun before the base year, it would take the landowner 20 years to get it fully established and thus be eligible for offsetting credits (521.2 tons). With 30 years remaining, the annual removals can result in a $CO_2$ offset of 409.8 tons (scenario 1-C). By 2050, the accountable $CO_2$ will amount to 931.0 tons. In contrast, without heeding to the additionality and permanence requirements, s/he might declare a total accumulated $CO_2$ offset of 2,576.6 tons—521.22 tons from standing trees and 68.38 tons a year from pulpwood and sawtimber produced over the 30 years (scenario 2-C).

explore how to increase the production of sawnwood and wood-based panels to extend the overall life of carbon stored in HWPs.

Due to these variations, coupled with the permanence principle, nonetheless, the potential of forest sector NbSs to mitigate climate change, albeit significant, may not be as great as some analysts have claimed (Hoel et al. 2014, Hou et al. 2020). In the case of intensively managed pine plantations in the US South, the outcome derived from the stand-level, Hartman-Faustmann framework with a neglect of the basic accounting principles and HWPs could exaggerate the amount of carbon credits a landowner may take by a factor of at least 2.76. On the other hand, while adding carbon to the valuation or raising carbon price has a limited impact on the optimal rotation age, the effect on net revenue is considerable. Thus, economic analysts must form a more balanced and consistent view of the influence of carbon price on management outcomes. In addition, unless ecosystem services other than timber and carbon become prominent, the rationale for maintaining certain FES over a longer term could be called into question because of the high capital costs of standing trees. I believe that these are important and timely findings, as the global community actively seeks NbSs to mitigate climate change (Qin et al. 2021, Fuss et al. 2018).

Similarly, complying with the Paris Agreement's requirements for carbon accounting does not imply that smallholders who have only a few stands of trees cannot participate in an emission offsetting and/or trading scheme or are not eligible for taking credits from their contributions. However, it does mean that there should be intermediary agencies who can aggregate the smallholders and bundle their individual properties, not only to make their participation feasible and trustworthy but also to reduce the impediments and costs entailed in monitoring, reporting, and verification (FAO 2016). Innovations in the brokerage of forest carbon investment and trading are thus needed, which in turn calls for a much-improved regulatory system of the current voluntary, as well as compliant, carbon markets (Nordhaus 2021).

The current study has examined the ramifications of different accounting principles and practices in the management of pine plantation forests in the US South, where data on stand growth and yield, output prices, and input costs are not just available but also of higher quality. Future research should extend this type of work to other FESs, such as naturally regenerated hardwood or mixed-species boreal forests in northern North America and Europe. Also, it is worthwhile to capture the status and dynamics of other FES carbon pools, such as soil. Further, I did not consider the possibility of generating biofuels and/or bio-chemicals from woody materials. In view of their promises to replace fossil fuels and/or non-renewable construction materials (Jonsson et al. 2021, Fuss et al. 2018), it is yet another important opportunity for future research. In these endeavors, though, more field experiments of carbon sequestration by different FESs and storage in HWPs are warranted, given that much of the current work on carbon accounting and credit taking is based on the Tier I parameters of the IPCC (2006), which are relatively crude.

Finally, as elaborated in Chapter 3, other than micro-level efforts of forest carbon accounting, more comprehensive assessments of the forest sector's potential in carbon emission reduction and/or removal have been done at the regional or national level (Lemprière et al. 2013, Baker et al. 2019). There also exist alternative frameworks, and their outcomes seem to deviate as well. It is thus necessary to investigate their appropriateness considering the Paris Agreement principles for carbon accounting and the relevant considerations of forest ecology and economics.

## Note

1 A slightly different version of this chapter has recently appeared in *PLOS Climate* (Sun et al. 2023).

## References

Asante, P., Armstrong, G.W. 2012. Optimal forest harvest age considering carbon sequestration in multiple carbon pools: A comparative statics analysis. *Journal of Forest Economics* 18 (2): 145–156.

Austin, K.G., Baker, J.S., Sohngen, B., Wade, C.M., Daigneault, A., Ohrel, S.B., Ragnauth, S., Bean, A. 2020. The economic costs of planting, preserving, and managing the world's forests to mitigate climate change. *Nature Communications* 11 (1): 1–9.

Baker, J.S., Forsell, N., Latta, G., Sohngen, B. 2019. State of the art methods to project forest carbon stocks. *Journal of Forest Economics* 34: 1–5.

Bowes, M.D., Krutilla, J.V. 1989. *Multiple-Use Management: The Economics of Public Forestlands*, 1st ed. Washington, DC: Resources for the Future.

Ekholm, T. 2016. Optimal forest rotation age under efficient climate change mitigation. *Forest Policy and Economics* 62: 62–68.

Food and Agriculture Organization of the United Nations (FAO). 2016. *Forestry for a Low-Carbon Future: Integrating Forests and Wood Products in Climate Change Strategies* (FAO Forestry Paper 177). Rome, Italy.

Fuss, S., Lamb, W.F., Callaghan, M.W., Hilaire, J., et al. 2018. Negative emissions—part 2: Costs, potentials and side effects. *Environmental Research Letters* 13 (6). https://doi.org/10.1088/1748-9326/aabf9f.

Galik, C., Murray, B.C., Mitchell, S., Cottle, P. 2016. Alternative approaches for addressing non-permanence in carbon projects: An application to afforestation and reforestation under the Clean Development Mechanism. *Mitigation and Adaptation Strategies for Global Change* 21: 101–118.

Gutrich, J., Howarth, R.B. 2007. Carbon sequestration and the optimal management of New Hampshire timber stands. *Ecological Economics* 62 (3-4): 441–450.

Hartman, R. 1976. The harvesting decision when a standing forest has value. *Economic Inquiry* 14 (1): 52–58.

Hoel, M., Holtsmark, B., Holtsmark, K. 2014. Faustmann and the climate. *Journal of Forest Economics* 20 (2): 192–210.

Holtsmark, B., Hoel, M., Holtsmark, K. 2013. Optimal harvest age considering multiple carbon pools—A comment. *Journal of Forest Economics* 19 (1): 87–95.

Hou, G.L., Delang, C.O., Lu, X.X., Olschewski, R. 2020. Optimizing rotation periods of forest plantations: The effects of carbon accounting regimes. *Forest Policy and Economics* 118 (102263).

Hyde, W.F. 2012. *The Global Economics of Forestry*, 1st ed. New York: RFF Press/Routledge.

International Panel on Climate Change (IPCC). 2006. *Guidelines for National Greenhouse Gas Inventories*. Available at: www.ipcc.ch/report/2006-ipcc-guidelines-for-national-greenhouse-gas-inventories/.

International Panel on Climate Change (IPCC). 2019. *2019 Refinement to the 2006 IPCC Guidelines for National Greenhouse Gas Inventories*. Available at: www.ipcc.ch/report/2019-refinement-to-the-2006-ipcc-guidelines-for-nationalgreenhouse-gas-inventories/.

International Panel on Climate Change (IPCC). 2020. *Climate Change and Land: Summary for Policymakers*. Available at: www.ipcc.ch/srccl/.

Irland, L.C., Sample, V.A. 1995. *Ecosystem Management in Northern Forests: Potential Role in Managing the Carbon Cycles*. Washington, DC: Forest Policy Center (Pinchot Institute Paper).

Jonsson, R., Rinaldi, F., Pilli, R., Fiorese, G., Hurmekoski, E., Cazzaniga, N., Robert, N., Camia, A. 2021. Boosting the EU forest-based bioeconomy: Market, climate, and employment impacts. *Technological Forecasting & Social Change* 163 (120478).

Lemprière, T.C., Kurz, W.A., Hogg, E.H., Schmoll, C., Rampley, G.J., Yemshanov, D., McKenney, D.W., Gilsenan, R., Beatch, A., Blain, D., Bhatti, J.S., Krcmar, E. 2013. Canadian boreal forests and climate change mitigation. *Environmental Reviews* 21: 293–321.

Li, X.Y., Lu, G., Yin, R.S. 2020. Research trends: Adding a profit function to forest economics. *Forest Policy and Economics* 113 (102133).

Nordhaus, W.D. 2021. *The Spirit of Green: The Collisions and Contagions in a Crowded World*. Princeton, New Jersey: Princeton University Press.

Ou, Y., Iyer, G., Clarke, L., Edmonds, J. et al. 2021. Can updated climate pledges limit warming well below 2°C? *Science* 6568: 693–695.

Pienaar, V., Rheney, J.W. 1995. Modeling stand level growth and yield response to silvacultural treatments. *Forest Science* 41 (3): 629–638.

Qin, Z., Deng, X., Griscom, B., Huang, Y., Li, T., Smith, P., Yuan, W., Zhang, W. 2021. Natural climate solutions for China: The last mile to carbon neutrality. *Adv. Atmos. Sci.* 38: 889–895.

Rennert, K., Prest, B.C., Pizer, W.A., Newell, R.G., Anthoff, D., Kingdon, C., Rennels, L., Cooke, R., Raftery, A.F., Ševčíková, H., Errickson, F. 2021. *The Social Cost of Carbon: Advances in Long-Term Probabilistic Projections of Population, GDP, Emissions, and Discount Rates* (Brookings Papers on Economic Activity).

Roe, S., Streck, C., Obersteiner, M., Frank, S., Griscom, B., Drouet, L., et al. 2019. Contribution of the land sector to a 1.5 °C world. *Nature Climate Change* 9: 817–828. doi:10.1038/s41558-019-0591-9.

Seymour, F., Busch J. 2016. *Why Forests? Why Now? The Science, Economics, and Politics of Tropical Forests and Climate Change*, 1st ed. Washington, DC: Brookings Institution Press.

Sohngen, B., Parisa, Z., Marland, E., Marland, G., Jenkins, J. 2022. The time value of carbon storage. *Forest Policy and Economics* 144 (102840).

Sun, H.Y., Midkiff, D., Yin, R.S. 2023. Different outcomes of alternative forest carbon accounting approaches at the local level. *PLOS Climate* 2 (5): e0000191. https://doi.org/10.1371/journal.pclm.0000191.

Swallow, S.K., Parks, P.J., Wear, D.N. 1990. Policy-relevant nonconvexities in the production of multiple forest benefits. *Journal of Environmental Economics and Management* 19: 264–280.

United Nations Framework Convention on Climate Change (UNFCCC) Secretariat. 2020. *Reference Manual for the Enhanced Transparency Framework under the Paris Agreement*. Bonn, Germany.

Van der Gaast, W., Sikkema, R., Vohrer, M. 2018. The contribution of forest carbon credit projects to addressing the climate change challenge. *Climate Policy* 18 (1): 42–48.

Van Kooten, G.C. 2018. The challenge of mitigating climate change through forestry activities: What are the rules of the game? *Ecological Economics* 146: 35–43.

Van Kooten, G.C., Binkley, C.S., Delcourt, G. 1995. Effect of carbon taxes and subsidies on optimal forest rotation age and supply of carbon services. *American Journal of Agricultural Economics* 77 (2): 365–374.

Wang, Y.F., Li, L.Y., Yin, R.S. 2021. A primer on forest carbon policy and economics under the Paris Agreement. *Forest Policy and Economics* 132 (102595).

Wear, D.N., Coulston, J.W. 2019. Specifying forest sector models for forest carbon projections. *Journal of Forest Economics* 34: 73–97.

Yin, R.S. 1997. An alternative approach to forest investment assessment. *Can. J. For. Res.* 27 (12): 2072–2078.

Yin, R.S., Newman, D.H. 1997. Long run timber supply and the economics of timber production. *Forest Science* 43: 113–120.

Yin, R.S., Pienaar, L.E., Aronow, M.E. 1998. The productivity and profitability of fiber farming. *Journal of Forestry* 96 (11): 13–18.

# 5 The forest sector's potential role in national decarbonization

## China as an example

### Introduction

China, the world's largest greenhouse gas (GHG) emitter, has acted on climate change by committing to a drastic reduction of its emissions beginning in this decade and has raised its ambition by aiming to reach net-zero emissions before 2060 (Mallapaty 2020).[1] Also, the Chinese government has reiterated its intention to increase the country's forest stock volume by six billion cubic meters ($10^9 m^3$) during 2006–2030 to partially offset its carbon emissions (Xinhua 2020). Thereafter, forest ecosystems (FESs) as well as harvested wood products (HWPs), are expected to play an even larger part in decarbonizing the economy (Energy Foundation China 2020).

FESs in China have greatly expanded in the last decades. Official inventory data (NFGA 2019) indicate that the country's forest area was 176.06 million hectares ($10^6 ha$) in 2005, rising to $220.44 \times 10^6 ha$ in 2018; similarly, stock volumes rose from $12.96 \times 10^9 m^3$ to $17.56 \times 10^9 m^3$ from 2005 to 2018. Meanwhile, China has become the second largest consuming country for log, lumber, wood panel, paper, and other wood products, with the total demand for commercial timber reaching $556.75 \times 10^6 m^3$ (in roundwood equivalent) in 2018 (Ke et al. 2020). A domestic timber supply of $364.88 \times 10^6 m^3$ for the same year implies that the remaining $191.87 \times 10^6 m^3$ had to be imported from overseas (Ke et al. 2020).

A question of broad international interest is thus: How significant a role can China's forest sector play in achieving the country's goal of net-zero emissions? To tackle this question, it is essential to project China's carbon sequestration by FESs and storage in HWPs, in addition to its carbon emissions (IPCC 2006). Recent research advances (Duan et al. 2021, Piao et al. 2009, Jiang et al. 2016, Johnston and Radeloff 2019), coupled with the accumulated information in these areas, makes it possible to formally quantify the role that China's forest sector has played in offsetting its carbon emissions so far and the role it will likely play in the future. Moreover, it is crucial to put China's potential achievements in emissions reduction and removal (ER&R) into a coherent accounting framework to conform to international requirements. In this way, ER&R can be counted toward fulfilling its Nationally

DOI: 10.4324/9781003436652-6

*The forest sector's potential role in national decarbonization* 79

determined contribution (NDC). Of particular relevance to forest carbon accounting are the principles of additionality, permanence, and no leakage measured against an established baseline (UNFCCC 2020, FAO 2016).

In this chapter, therefore, I assess the role of China's FESs and HWPs in carbon emissions offsetting by integrating the projected outcomes of ER&R into alternative scenarios within a harmonized accounting framework. In addition to confirming the significance of forest-based solutions to the country's efforts of climate change mitigation, this study can serve as an example of carbon accounting, offset crediting, and tracking toward meeting the targets specified by countries in their NDCs.[2]

## Principles and methods

In this section, I overview the principles of carbon accounting under the Paris Agreement and describe the methods used in estimating and projecting China's ER&R over time.

### *Accounting principles*

The Paris Agreement requires that any ER&R come from human actions (IPCC 2019), with an exclusion of the effects of natural causes. That is, as elaborated in Chapter 1, a mitigation accomplishment can be accounted for only when it represents a reduction in anthropogenic emissions relative to a mutually agreed *baseline* or reference level (Grassi et al. 2021, Ogle and Kurz 2021). Thus, a key first step in carbon accounting is defining a country's baseline, against which the *additionality* of its emissions or removals can be determined. In net-net accounting, total net ER&R in the base year are subtracted from total net ER&R in the future accounting period to find the total additional number of credits or debits resulting from these actions (UNFCCC 2020). This approach is well suited to China.[3]

Since almost all carbon captured by trees is eventually emitted back to the atmosphere after the trees are harvested or lost to natural mortality or disturbances, any forest-based action can lead to non-permanence if not well designed. *Permanence* mandates that carbon sequestered and stored that has entered an accounting system will not be released over a sufficiently long period (Sedjo and Sohngen 2012, Galik et al. 2016). The duration of an action should be commensurate with the life of an NDC. Moreover, in addition to changes in land use practices, the possibility of commodity trade between jurisdictions means that it is necessary to consider the carbon flow associated with the product flow because these situations can lead to carbon *leakage* elsewhere, i.e., a displacement of the emissions (FAO 2016).

In line with China's NDC, my accounting takes its 2005 ER&R estimates as the baseline levels, and my projections run up to 2060. Also, a consensus of the science community is that the country's carbon removals by its FESs will have all come from human actions, be they afforestation and reforestation driven by ecological restoration efforts, commercial plantation development, conservation of protected areas, or the management of existing forests (Fang et al. 2001, Yin 2009, Jiang et al. 2016). Further, I argue that for China, carbon sequestered by FESs and stored in HWPs will be sustained at the national level, conforming to the permanence condition. Meanwhile, given the country's high dependence on imported timber resources to fulfill its needs for domestic consumption and re-exporting of finished products (Ke et al. 2020), the likely carbon deficit must be accounted for to avoid any leakage. Finally, I focus on $CO_2$ ER&R in this assessment, considering its relevance to forest-based mitigation and the challenge in projecting other GHGs (e.g., $CH_4$, $NO_2$) (Duan et al. 2021).

***Estimation methods***

Carbon sequestered by a forest is distributed in multiple pools, including above-ground biomass, below-ground biomass, and soil (Fang et al. 2001). A certain amount of the above-ground biomass may become merchantable products; once biomass is removed for timber or fiber, a portion of the carbon will be stored in HWPs, with the rest reemitted back to the atmosphere. Also, HWPs have different service lives, with solid wood products (e. g., housing structures and furniture) being much more durable than those of reconstituted wood fibers (e.g., paper and paperboard) (Fuss et al. 2018). As such, the actual amount of carbon stored in HWPs must be assessed properly. In any case, the cumulative removal of $CO_2$ by FESs and HWPs is the sum of carbon sequestered in biomass and soil and stored in HWPs over a given period. The ratio of this cumulative removal over the cumulative emission during the same period reflects the offsetting potential for forest sector actions.

To save space, I summarize the data sources and modeling alternatives used in estimating and projecting ER&R below, with detailed steps and assumptions put in Appendix I. The historical statistics of China's carbon emissions (2005–2019) were obtained from *Our World in Data* [4]. Beginning from 2020, its carbon emissions were projected by using three of the integrated assessment models contained in Duan et al. (2021), including the Global Change Analysis Model operated by Tsinghua University (GCAM-TU), the Integrated Policy Assessment model of China (IPAC), and the World Induced Technical Change Hybrid model (WITCH).

The relevant National Forest Inventories (NFIs: 7th for 2004–2008; 8th for 2009–2013; and 9th for 2014–2018) were used to determine changes in the historical forest condition until 2018, with data for the intermittent years being interpolated. Post-2018 projections of the aggregate stock volume were based on the expansion of forest area and the growth of unit stocking volume in reference to the targets set by the National Forest and Grassland Administration. Two alternative cases—one conventional and the other aggressive growth—were considered.

The historical statistics for quantifying carbon stored in HWPs are available from the *FAOSTAT* database, which provides country-level data on the production, import, and export of wood products[5]. The carbon contents contained in sawnwood, wood-based panels, and paper and paperboard products were estimated using the stock-change approach recommended by the IPCC (2006), in which the trade-induced carbon flows in these products are captured and leakage avoided (FAO 2016). But before doing so, it is also necessary to make projections of the future market conditions for China. Fortunately, Johnston and Radeloff (2019) have done so.

## Building scenarios

Uncertainties exist in the projected carbon emission, FES sequestration, and HWP storage. As noted by Duan et al. (2021), the available emission space is affected by factors such as assumptions on present-day warming, the magnitude of non-carbon emissions, climate sensitivity, and the extent of mitigation activities. Likewise, FES sequestration and HWP storage are driven by social, economic, technological, and environmental forces (Johnston and Radeloff 2019). Consequently, deviations between what is projected and what will occur are inevitable. We could formulate a distribution around the projected values for each of the crucial variables to ensure the reliability of the assessed offsetting outcome, but this would be a complicated exercise beyond the scope of this study. Otherwise, I could construct alternative scenarios, similar to the Shared Socioeconomic Pathways used in the literature (e.g., IPCC 2022, Johnston and Radeloff 2019), based on my professional experience and knowledge to deal with the uncertainties embedded in my assessment. I chose the latter.

I began by building two basic classes of scenarios—one conservative and the other aggressive—to gauge the possible range of the projected outcomes and the ultimate effects of an integrated assessment of ER&R. The former combined higher projected emissions with lower FES sequestration and HWP storage, which should thus have a greater likelihood of realization. The latter features an opposite situation, lower projected emissions coupled with more aggressive FES sequestration and HWP storage, which would be harder to achieve. Further, I built the two classes of scenarios in a step-by-step manner by examining:

1 FES biomass sequestration only, with the exclusion of soil sequestration and HWP storage;
2 FES biomass sequestration and HWP storage, with the exclusion of soil sequestration; and
3 FES biomass and soil sequestration and HWP storage.

Together, these steps resulted in a total of six concrete scenarios, which I hope will defuse concerns about the trustworthiness of my findings.

82 *The forest sector's potential role in national decarbonization*

## Results

Here, I proceed from the projected carbon emission, sequestration, and storage to the estimated offsets under different scenarios. More details can be found in the Supplementary Material of Hou and Yin (2022).

### *Emissions*

Two trends emerged from the projected outcomes. First, China's carbon emission would peak soon after 2020, at a level of ≤ 2.950Pg, as projected by the IPAC model. Given its sizable emission drop caused by the pandemic-induced economic slowdown and the announcement of net-zero emissions by 2060, this earlier peaking looks more likely than otherwise. Second, the largest emissions reductions will have to happen during this and the next decade (Duan et al. 2021).

As shown in Figure 5.1, China's carbon emission was 1.574Pg in 2005 and reached 2.774Pg in 2019. The GCAM-TU projected that it would decrease, respectively, to 2.141Pg in 2030, 1.136Pg in 2040, and 0.300Pg in 2050. The cumulative emissions would be 36.792Pg from 2006 to 2020, 23.776Pg during 2021–2030, 15.999Pg during 2031–2040, 6.130Pg during 2041–2050, and 0.828Pg during 2051–2060. Altogether, the aggregate carbon emissions would be 83.526Pg during 2006–2060 (see Table 5.1). Similarly, the IPAC model projected that carbon emissions would decrease to 2.264Pg in 2030, 1.366Pg in 2040, 0.322Pg in 2050, before reaching zero in 2060. The cumulative emissions would be 25.672Pg during 2021–2030, 17.323Pg during 2031–2040, 7.417Pg during 2041–2050, and 0.981Pg during 2051–2060. The total carbon emissions would thus be 88.425Pg during 2006–2060.

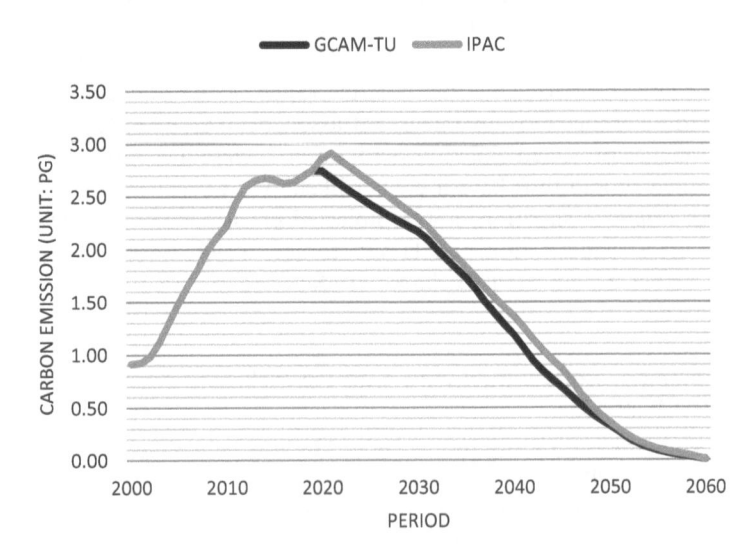

*Figure 5.1* Alternative trajectories of carbon emissions projected by the GCAM-TU and the IPAC model

*Table 5.1* China's cumulative C emissions (unit: petagrams)

| Time period | GCAM-TU | WITCH | IPAC |
|---|---|---|---|
| 2006–2020 | 36.792 | 37.015 | 37.031 |
| 2021–2030 | 23.776 | 18.072 | 25.672 |
| 2031–2040 | 15.999 | 9.620 | 17.323 |
| 2041–2050 | 6.130 | 6.207 | 7.417 |
| 2051–2060 | 0.829 | 3.345 | 1.104 |
| 2006–2060 | 83.526 | 74.259 | 88.547 |

*Notes*:
1) The three sets of emission projections were taken from Duan et al. (2021), with GCAM-TU, WITCH, and IPAC representing the Global Change Analysis Model of Tsinghua University, the World Induced Technical Change Hybrid model, and the Integrated Policy Assessment model of China.
2) 2005 is the base year for assessing China's carbon emission reduction, and 2060 is the time when it will reach net zero.

## Sequestration

Figure 5.2 illustrates the trajectories of forest area and stock volume over time. Under the conventional case, it would reach 239.98 × 10⁶ha and 20.674 × 10⁹m³ by 2030, and 249.58 × 10⁶ha and 23.242 × 10⁹m³ by 2040. Thereafter, the forest area would become stabilized, while the stock volume could increase to 25.402 × 10⁹m³ by 2050 and, finally, 27.772 × 10⁹m³ by 2060. Consequently, the total forest area and stock volume would increase by 66.451 × 10⁶ha and 14.816 × 10⁹m³. Compared to the 2005 level, forest stock volume in 2020 would gain by 5.063 × 10⁹m³; similarly, the stock volume increase during 2006–2030 would reach 7.709 × 10⁹m³, much larger that the latest official target of 6 × 10⁹m³. This means that the added forest biomass and soil C during 2006–2020 amounted to 2.589Pg and 1.275Pg, leading to a total FES sequestration of 3.864Pg as summarized in Table 5.2-A. Likewise, the additional biomass and soil carbon would reach 7.573Pg and 3.730Pg for 2006–2060, resulting in an overall sequestration of 11.304Pg.

While the above estimates are impressive, recent studies suggest that China can do even better in boosting its forest growth by implementing extensive thinning and other stand treatments (Li et al. 2020, Hou et al. 2019). Under the accelerated case, the unit stock would increase to 88.14m³/ha by 2030, 96.88m³/ha by 2040, 107.02m³/ha by 2050, and 118.22m³/ha by 2060. Compared to the conventional case, the last figure reflects a gain of 6.25% over the four decades (2021–2060). In correspondence, the stock volume gain would amount to 3.128 × 10⁹m³ during 2021–2030, 3.028 × 10⁹m³ during 2031–2040, 2.530 × 10⁹m³ during 2041–2050, and 2.794 × 10⁹m³ during 2051–2060. The aggregate gain in forest stock volume would be 16.542 × 10⁹m³ over 2006–2060 (see Table 5.2-B), leading to biomass and soil carbon additions of 8.460Pg and 4.167Pg, respectively.

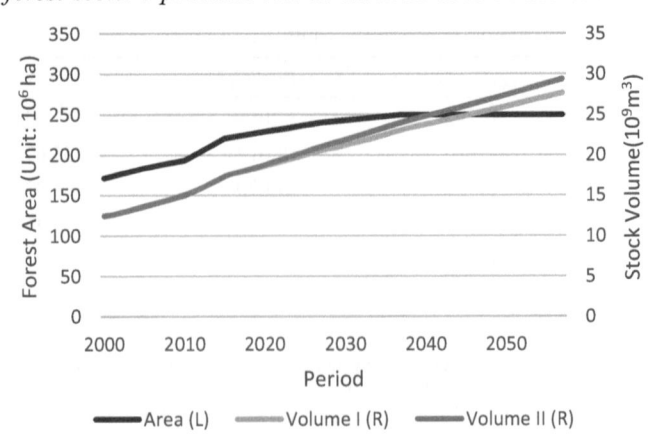

84  *The forest sector's potential role in national decarbonization*

*Figure 5.2* Alternative trajectories of projected forest area and stock volume under the conventional case (I) and the accelerated case (II)

*Table 5.2* China's C sequestration by forest ecosystems

**A. The conventional case**

| Period | Forest condition change ||| Cumulative sequestration |||
|---|---|---|---|---|---|---|
| | Area ($10^6$ ha) | Unit stock ($m^3$/ha) | Total stock ($10^9$ $m^3$) | Biomass (Pg) | Soil (Pg) | Ecosystem (Pg) |
| 2006–2020 | 40.536 | 9.808 | 5.063 | 2.589 | 1.275 | 3.864 |
| 2021–2030 | 16.316 | 5.561 | 2.649 | 1.355 | 0.667 | 2.022 |
| 2031–2040 | 9.599 | 6.977 | 2.568 | 1.313 | 0.647 | 1.960 |
| 2041–2050 | 0.000 | 8.670 | 2.164 | 1.107 | 0.545 | 1.652 |
| 2051–2060 | 0.000 | 9.478 | 2.365 | 1.210 | 0.596 | 1.805 |
| 2006–2060 | 66.451 | 37.688 | 14.816 | 7.573 | 3.730 | 11.304 |

**B. The accelerated case**

| Period | Forest condition change ||| Cumulative sequestration |||
|---|---|---|---|---|---|---|
| | Area ($10^6$ ha) | Unit stock ($m^3$/ha) | Total stock ($10^9$ $m^3$/ha) | Biomass (Pg) | Soil (Pg) | Ecosystem (Pg) |
| 2006–2020 | 40.536 | 9.808 | 5.063 | 2.589 | 1.275 | 3.864 |
| 2021–2030 | 16.316 | 7.554 | 3.128 | 1.599 | 0.788 | 2.387 |
| 2031–2040 | 9.599 | 8.741 | 3.028 | 1.548 | 0.763 | 2.311 |
| 2041–2050 | 0.000 | 10.136 | 2.530 | 1.294 | 0.637 | 1.931 |
| 2051–2060 | 0.000 | 11.197 | 2.794 | 1.429 | 0.704 | 2.133 |
| 2006–2060 | 66.451 | 47.436 | 16.542 | 8.460 | 4.167 | 12.626 |

*Notes*:
1) See Section 2 and the Appendix for data sources and estimation assumptions and procedures. More details can be found in the Supplementary Material.
2) Forest area changes over time are the same for both cases.
3) Driven by more intensive management practices, the accelerating annual growth rates were assumed to be 0.90% during this decade, 0.95% during the next decade, and 1.00% thereafter.

## Storage

The amount of carbon storage in HWPs would peak around 2030 and decline afterward, implying that the increasing trend of China's consumption of various wood products could only be maintained for one more decade (Johnston and Radeloff 2019). However, as its economic growth rate slows, some of the accumulated HWPs during the rapid growth period would move out of service. If not managed properly, the decaying HWPs could cause a release of the stored carbon and thus a downward drift in future annual net storage. Ideally, these post-service HWPs could be further used to avoid some additional amount of annual carbon emission through substituting fossil fuels, or their derivative products, in the manufacturing of refurbished commodities or the generation of bioenergy (Jonsson et al. 2021, Fuss et al. 2018). Figure 5.3 highlights this trend and the relatively small deviations between the two different pathways, with SSP4 delivering the lowest carbon storage change (1.709Pg) and SSP5 the highest (1.899Pg) over 2006–2060. On average, these pathways would imply a small net annual removal from 0.031Pg to 0.0345Pg (see Table 5.3).

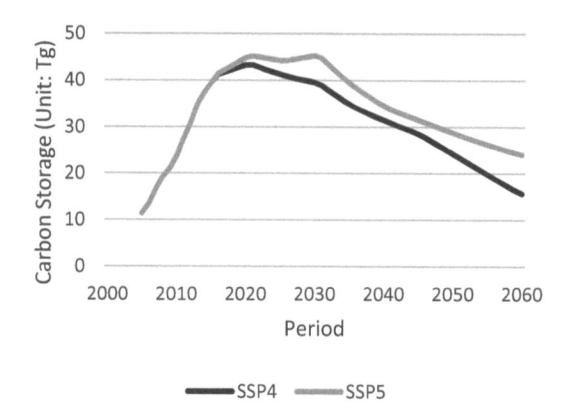

*Figure 5.3* Alternative pathways (SSPs) of C storage in HWPs

*Table 5.3* China's carbon storage in harvested wood products (unit: petagrams)

| Period | SSP4 | SSP5 |
| --- | --- | --- |
| 2006–2020 | 0.491 | 0.496 |
| 2021–2030 | 0.410 | 0.447 |
| 2031–2040 | 0.344 | 0.387 |
| 2041–2050 | 0.275 | 0.310 |
| 2051–2060 | 0.189 | 0.258 |
| 2006–2060 | 1.709 | 1.899 |

*Notes:*
SSP4 and SSP5 represent two of the extreme shared socioeconomic pathways derived from the projections made by Johnston and Radeloff (2019). More details can be found in the Supplementary Material.

## Offsetting

The conservative class of scenarios was based on a combination of higher projected carbon emissions generated by the IPAC model, and lower FES sequestration under Case I and lower HWP storage associated with SSP4. Specifically, during 2006–2020, the sum of carbon stored in HWPs amounted to 0.491Pg, carbon sequestered by forest biomass and soil were 2.589Pg and 1.275Pg, respectively, while the cumulative emission was 37.031Pg (Table 5.4-A). If only FES biomass carbon is considered, it gives an offsetting ratio of 6.99%; if carbon sequestered by FES biomass and stored in HWPs are included, the offsetting ratio rises to 8.32%. Otherwise, if carbon sequestered by FES biomass and soil and stored in HWPs are all included, the offsetting ratio reaches 11.76%. Put differently, the amount of carbon sequestered by FESs and stored in HWPs would be even greater than the total emissions of the transportation sector (Energy Foundation China 2020). During this decade, China's FESs could sequester another 2.022Pg of carbon, which, when combined with the HWP carbon storage of 0.410Pg, would offset 9.47% of the carbon emissions.

Before 2040, China's forest carbon sequestration relies primarily on area expansion, with increased volume growth accounting for a smaller portion (Li et al. 2016). During 2031–2040, FES carbon would add another 1.960Pg; with an increase of 0.344Pg of carbon stored in HWPs, the total offset could reach 2.304Pg. Due to the substantially declined decadal emissions (17.323Pg), this would yield an offsetting ratio of 13.30%. Later, carbon

Table 5.4 A summary of the historical and projected carbon emissions and removals (Pg), and offset ratios (%)

### A. The conservative class of scenarios

| Time | Emission | Biomass C | Soil C | HWP C | FS $C_a$ | FS $C_b$ | Offset I | Offset II | Offset III |
|---|---|---|---|---|---|---|---|---|---|
| 2006–2020 | 37.031 | 2.589 | 1.275 | 0.491 | 3.080 | 4.355 | 6.99 | 8.32 | 11.76 |
| 2021–2030 | 25.672 | 1.355 | 0.667 | 0.410 | 1.765 | 2.432 | 5.28 | 6.87 | 9.47 |
| 2031–2040 | 17.323 | 1.313 | 0.647 | 0.344 | 1.657 | 2.304 | 7.58 | 9.57 | 13.30 |
| 2041–2050 | 7.417 | 1.107 | 0.545 | 0.275 | 1.381 | 1.926 | 14.92 | 18.63 | 25.97 |
| 2051–2060 | 0.981 | 1.210 | 0.596 | 0.189 | 1.399 | 1.995 | 123.31 | 142.60 | 222.63 |
| 2006–2060 | 88.425 | 7.574 | 3.730 | 1.709 | 9.282 | 13.013 | 8.57 | 10.50 | 14.72 |

Notes:
1) Emissions, projected by the IPAC model (see Table 5.1), represent a case on the higher end.
2) Biomass C (carbon) is the forest biomass carbon derived using the biomass expansion and C conversion factors, corresponding to the conventional case (see Table 5.2).
3) Soil C is derived on the basis of a 33% share of the forest ecosystem C.
4) HWP C is C stored in harvested wood products under SSP4 (see Table 5.3).
5) FS C is the forest-sector total C uptake—a sum of forest ecosystem and HWP C, and FS $C_a$ and FS $C_b$ differ in terms of whether soil C is included.
6) Offset I, II, and III are the offsetting ratios corresponding to forest biomass C only, forest biomass and HWP C, and forest ecosystem and HWP C, respectively.

*Table 5.4 continued*

B. The optimistic class of scenarios

| Time | Emission | Biomass C | Soil C | HWP C | FS $C_a$ | FS $C_b$ | Offset I | Offset II | Offset III |
|---|---|---|---|---|---|---|---|---|---|
| 2006–2020 | 36.792 | 2.592 | 1.277 | 0.496 | 3.088 | 4.365 | 7.037 | 8.386 | 11.85 |
| 2021–2030 | 23.776 | 1.599 | 0.788 | 0.447 | 2.046 | 2.834 | 6.727 | 8.735 | 12.05 |
| 2031–2040 | 15.999 | 1.548 | 0.763 | 0.387 | 1.935 | 2.311 | 9.678 | 12.094 | 16.86 |
| 2041–2050 | 6.130 | 1.294 | 0.637 | 0.310 | 1.604 | 1.931 | 21.104 | 26.160 | 36.56 |
| 2051–2060 | 0.829 | 1.429 | 0.704 | 0.258 | 1.687 | 2.133 | 172.382 | 203.535 | 288.44 |
| 2006–2060 | 83.526 | 8.463 | 4.168 | 1.899 | 9.286 | 14.531 | 10.132 | 12.405 | 17.39 |

*Notes*:
1) Emissions, projected by the GCAM-TU (see Table 5.1), represent a case on the lower end.
2) Biomass carbon (C) is the forest biomass C derived using the biomass expansion and C conversion factors, corresponding to the accelerated case (see Table 5.2).
3) Soil C is derived on the basis of a 33% share of the forest ecosystem C.
4) HWP C is C stored in harvested wood products under SSP5 (see Table 5.3).
5) FS C is the forest-sector total C uptake—a sum of forest ecosystem and HWP C, and FS $C_a$ and FS $C_b$ differ in terms of whether soil C is included.
6) Offset I, II, and III are the offsetting ratios corresponding to forest biomass C only, forest biomass and HWP C, and forest ecosystem and HWP C, respectively.

sequestered by FES would rely exclusively on forest growth. The amounts of carbon sequestration and storage would be 1.652Pg and 0.275Pg during 2041–2050, and 1.805Pg and 0.189Pg during 2051–2060, leading to an offsetting ratio of 25.97% and 222.63% respectively. From 2006 to 2060, FES biomass and soil carbon sequestration would be 7.573Pg and 3.730Pg, and carbon storage in HWPs would be 1.709Pg. Accordingly, forest biomass carbon alone gives an offsetting ratio of 8.57%; carbon sequestered by FESs in biomass and stored in HWPs yields an offsetting ratio of 10.50%; and carbon sequestered by FES biomass and soil and stored in HWPs together results in an offsetting ratio of 14.72%.

If appropriate and prompt stand treatment activities can be executed, I expect that during 2006–2060, carbon captured by FES biomass and soil would reach, respectively, 8.463Pg and 4.168Pg under the accelerated-growth case (Table 5.4-B). Coupled with more optimistic trajectories of carbon emission (83.526Pg per the GCAM-TU) and HWP storage (1.899Pg under SSP5), the total forest sector carbon removal could deliver an offset ratio of 17.39%. Even if only carbon sequestered by FES biomass is considered, it would offset 10.13% of the emissions. Similarly, carbon sequestered by FES biomass and stored in HWPs would offset 12.41% of the emissions.

## Discussion

As countries begin to fulfill their NDCs and further raise their ambitions to combat climate change, it is necessary and timely to assess the significance of their different ER&R actions. Based on my interpretation of the principles of carbon accounting for ER&R under the Paris Agreement, this assessment has shown for the first time that China's FESs and HWPs have played and will continue to play an important role in its drive to carbon emission reduction and neutrality even under the conservative mitigation scenarios.

My analytic strategy is similar to the one adopted in an earlier study (Yin et al. 2010), whose feasibility has been validated by the outcomes of this assessment regarding China's carbon emitted and sequestered in its forest biomass over the period of 2006–2020. That study projected that the cumulative carbon emissions during that period would fall in the range of 32.7–34.4 × $10^9$Pg. It also predicted that if the unit stocking level of its forests could be raised by 10 m$^3$/ha during 2006–2020, along with an area expansion of 40 × $10^6$ha, the stock volume gain would be 5.02 × $10^9$m$^3$, offsetting about 8.00% of the carbon emissions. Here, I have found that the actual gain of forest stock volume was 5.063 × $10^9$m$^3$, implying an offsetting ratio of 6.99%—slightly below what was estimated by Yin et al. (2010) due to the higher level of cumulative emissions (37.031Pg derived from the IPAC model) than what was predicted earlier.

Now, it is useful to recall the evolution of China's forest stock volume targets over time in its climate action planning. The original target set by the government in 2009 was to raise the country's forest stock level by only 1.3×$10^9$ m$^3$ for the period of 2006–2020 (Yin et al. 2010). Later in 2015, included in its first Nationally Determined Contribution was a target of increasing its forest stock volume by 4.5×$10^9$ m$^3$ for the period of 2006–2030 (NDRC 2015), which is 1.0×$10^9$ m$^3$ lower than what has been achieved. In raising its climate ambition in 2019, the government expressed its intention to increase the country's forest stock volume by 6.0×$10^9$ m$^3$ from 2006 to 2030 (Xinhua 2020). This assessment suggests that the stock volume of China's forests could grow by 7.7×$10^9$m$^3$ during that period. So, the government has been repeatedly conservative in setting its target for forest stock volume. Furthermore, China has not set targets for its forest sector post 2030, even though it is expected to play an even larger role (Energy Foundation China 2020).

In addition to elucidating the significance of negative emission technologies as reflected in FES sequestration and HWP storage (Fuss et al. 20198), I have argued that these actions are more cost effectiveness in achieving China's ER&R goals. Studies have indicated that the country's transformation of its energy infrastructure would entail extremely high costs (Energy Foundation China 2020). Duan et al. (2021) revealed that the models they reviewed gave China a social cost of carbon in 2050 between $315 and $2,240 per tCO$_2$ (in 2010 US dollars). In contrast, a recent case study in northeast China suggests that the carbon costs of timber plantations are in the range of $23.4 to 45.1 per tCO$_2$ (Li et al. 2021). This finding of low abatement costs of forest sector actions is partially confirmed by a national-level assessment of reforestation projects in pan-tropic countries by Busch et al. (2019).

Moreover, investing in thinning and other stand treatments can considerably improve the forest structure, quality, and productivity. As of 2018, China's average stocking level of regular, arbor forests (Li et al. 2016) was 95 m³/ha, that of plantation forests was 59.30 m³/ha, and the overall mean annual increment was 4.73 m³/ha. These indicators are markedly below their international counterparts, as well as China's forestland potential (Ke et al. 2020, Hou et al. 2019). Adopting more effective silvicultural practices will thus accelerate forest growth and the generation of more timber, fiber, and other non-timber forest products, which will lead to more materials available for the domestic wood products industry and bioenergy production, as well as greater amounts of carbon to be sequestered and other services provided (Hou et al. 2019). More consumption of domestic timber can in turn reduce China's ecological footprints overseas and increase carbon stored in HWPs (Ke et al. 2020, Hou et al. 2019). Together, these shifts will help China to advance toward meeting its national goals of climate change mitigation and ecological civilization, including the FES resilience and adaptability (Yin et al. 2010).

Indeed, China's forest sector could help the country neutralize its emissions before 2050 and achieve net-zero at least one decade earlier than what the government has envisioned. Under the more aggressive scenario outlined earlier, the added stock volume would total $3.128 \times 10^9$ m³ for 2021–2030, $3.028 \times 10^9$ m³ for 2031–2040, $2.530 \times 10^9$ m³ for 2041–2050, and $2.794 \times 10^9$ m³ for 2051–2060. The aggregate gain in forest stock volume would reach $16.549 \times 10^9$ m³ over 2006–2060. This stock volume increase would translate into forest biomass and soil carbon additions of 8.463Pg and 4.168Pg. The sum of these amounts of biomass and soil carbon sequestration is 1.322Pg larger than the basic case presented in the last section. Adding carbon stored in HWPs to the FES sequestration would yield an upper-bound offsetting ratio of 16.22%.

Therefore, it is imperative for China to adopt more effective silvicultural practices. By pursuing extensive thinning and other stand treatments, it will be able to not only create jobs and incomes for local workers but also increase forest growth and domestic timber supply. As stand structure and growth accelerate, the FES productivity and functionality will improve, leading to greater amounts of carbon to be sequestered and other services provided (Liu and Yin 2012). The current annual thinning and actively managed forest area is not only small and slow, but also primarily driven by government financing (NFGA 2019).

More and higher-quality timber and fiber resources derived from accelerated forest growth will generate more materials available for the domestic wood product and bioenergy industries (Liu and Yin 2012). Greater consumption of domestic wood and fiver resources can in turn reduce China's ecological footprint overseas and increase carbon stored in HWPs, especially if more sawnwood and wood panels can be manufactured and consumed, and more fossil fuels and their derivative products and non-renewable construction materials (e.g., cement and steel) can be substituted by wood-based biofuels, bio-energies, or biomaterials (IPCC 2006 & 2019). Further, post-service HWPs can be recycled and reused in manufacturing reconstituted commodities or making biofuels (Jonsson et al. 2021, Hou and Yin 2022). Together, these opportunities will go a long way toward meeting the national goals of CC mitigation and ecological civilization.

To improve the efficiency and productivity of its forestry, however, China must further restructure its institutional arrangements and public policies. In addition to continuing forest tenure and organizational reforms, one priority for the authorities to consider is whether it is wise or necessary to ban logging and other silvicultural operations in all the natural forests (Sun et al. 2016). Another challenge is to determine whether more than a half of the total forestland has to be designated exclusively for ecological purposes and thus deprive community collectives or individuals, who own over 40% of the designated forests (Liu et al. 2016), of their commercial management and use rights. Without adequate rights and incentives to manage the forest and dispose of and benefit from cultivated trees, it is hard to imagine that farmers and others will make the needed efforts to enhance forest growth and yields (Liu and Yin 2012).

Also, biofuels and other bio-products generated from wood materials have not been well developed or adequately documented in China (NFGA 2021). As such, I was unable to take them into account in this assessment. Moreover, in addition to leaving logging residuals on the ground, burning or dumping post-service HWPs leads to a release of the carbon contained in them. Instead, these materials should be used or reused in the generation of bioenergy and the manufacturing of refurbished or reconstituted commodities. Through substituting fossil fuels and their derivative products, or replacing non-renewable construction materials (e.g., cement and steel), this kind of undertaking will avoid a certain additional amount of the carbon emission (Jonsson et al. 2021, Fuss et al. 2018). Future research should be conducted as more wood materials become available and greater nature-based solutions to climate change become more practical.

Also, currently, once a forest is classified for environmental purposes, normal silvicultural activities are restricted substantially or even forbidden completely, which has disrupted the operations of many forest enterprises and individuals (Liu and Xia 2021). The idea of separating commercial uses arbitrarily from environmental uses not only complicates resource management but also hampers the realization of optimal FES services, including C sequestration (Liu et al. 2016). Likewise, the nationwide logging quota system is detrimental to operational effectiveness, and the industrial processes of wood products manufacturing should be upgraded and better integrated with resource bases to achieve more efficient utilization of the available resources and greater satisfaction in fulfilling societal needs for multiple ecosystem services, including, in addition to provisioning of wood and other products, sequestering and storing C, regulating watersheds, and offering tourism and recreational opportunities (Hou et al. 2019).

## Conclusions

My assessment shows that even if only forest biomass carbon is considered, it could have offset 6.99% of the cumulative emissions during 2006–2020. When including carbon sequestered by forest biomass and soil and stored in HWPs, these actions combined could offset 11.76% of China's cumulative carbon emissions during 2006–2020. As China's annual emissions pass their peak this decade, the proportion offset by FESs and HWPs will likely decrease below 10.00%, but they are expected to play a larger role later.

Overall, China's FESs and HWPs could offset 14.72% of its cumulative carbon emissions during 2006–2060 even under conservative assumptions. If more optimistic projections could materialize, carbon sequestered in FESs and stored in HWPs would offset as much as 17.39% of its cumulative carbon emissions in this period. Moreover, my assessment suggests that China's forest sector carbon sequestration and storage could more than offset its cumulative emissions during the 2050s, implying that the country would achieve carbon neutrality a decade earlier than what the government has envisioned. Certainly, we have high confidence in China's compliance with the permanence and leakage, as well as additionality, rules of the NDC accounting (FAO 2016, Wang et al. 2021). To accomplish the needed transformations, China must further restructure its institutional arrangements and policies to improve the efficiency and productivity of its forestry in general and forest management in particular (Liu and Yin 2012). Without active and judicious human interventions, however, passive protection may not be the most effective strategy of forest management in the long run.

Finally, to account for forest contributions to climate change mitigation adequately, China must also make efforts to capture every possible carbon pool of the FESs, including in biomass and soil, in urban trees and forests, and in trees outside of forests in rural, agroforestry systems. Some of these pools have been underestimated or completely ignored by the conventional, merchantable-timber-centered NFI system (Guo et al. 2014, Zeng et al. 2015, FAO 2002). If these omitted or underestimated carbon pools could be properly measured and monitored, they would further raise the amount of FES removals (Wang et al. 2021, Cook-Patton et al. 2021).

In addition to continuing the forest tenure and organizational reforms, one priority for the authorities to reconsider is whether it is wise or necessary to ban logging and other silvicultural operations in all the natural forests, including those naturally regenerated secondary forests. Time has come to reassess the sensibility of setting aside massive swaths of productive forestland, particularly in the northeast, under the Natural Forest Protection Program (Hou et al. 2019), which has left little room for local communities to engage in thinning and other management activities.

## Notes

1 In 2018, China's $CO_2$ and GHG emissions stood, respectively, at 10.1 billion tons (i.e., petagrams or Pg) and 12.7Pg (Energy Foundation China 2020).
2 An earlier version of this chapter has appeared in *Climate Policy* (Hou and Yin 2022).
3 In addition to crediting, a results-based payment scheme for REDD+ (reducing emissions from deforestation, forest degradation, and conservation of forest carbon stock) may be pursued by a developing country in international cooperation to generate carbon offsets, whereby the UNFCCC focuses on mitigation beyond

business as usual (FAO 2016). In this context, the actions that should count are those that reduce $CO_2$ emissions above and beyond what would occur in the absence of their execution—another way of determining additionality.
4 An open-access data source run by the University of Oxford (https://ourworldindata.org/).
5 An open-access database run by the FAO (www.fao.org/faostat/en/#home).

## References

Aldy, J., Pizer, W., Tavoni, M., Reis, L.A., Akimoto, K., Blanford, G., Carraro, C., Clarke, L E. et al. 2016. Economic tools to promote transparency and comparability in the Paris Agreement. *Nature Climate Change* 6: 1000–1004.

Bosetti, V., Carraro, C., Galeotti, M., Massetti, E., Tavoni, M. 2006. WITCH: A World Induced Technical Change Hybrid Model. *The Energy Journal* 27: 13–37.

Busch, J., Engelmann, J., Cook-Patton, S.C., Griscom, B.W., Kroeger, T., Possingham, H., Shaymsundar, P. 2019. Potential for low-cost carbon dioxide removal through tropical reforestation. *Nature Climate Change* 9: 463–466.

Choumert, J., Motel, P.C., Dakpo, H.K. 2013. Is the Environmental Kuznets Curve for deforestation a threatened theory? A meta-analysis of the literature. *Ecological Economics* 90: 19–28.

Cook-Patton, S.C., Shoch, D., & Ellis, P.W. (2021). Dynamic global monitoring needed to use restoration of forest cover as a climate solution. *Nature Climate Change*, 11, 366–368.

Duan, H., Zhou, S., Jiang, K., Bertram, C., Harmsen, M., Kriegler, E., van Vuuren D.P. et al. 2021. Assessing China's efforts to pursue the 1.5°C warming limit. *Science* 372: 378–385.

Energy Foundation China. 2020. *Synthesis Report 2020 on China's Carbon Neutrality: China's New Growth Pathway: from the 14th Five Year Plan to Carbon Neutrality.* Available at https://www.efchina.org/Reports-en/report-lceg-20201210-en.

Fang, J.Y., Chen, A., Peng, C., Zhao, S., Ci, L. 2001. Changes in forest biomass carbon storage in China between 1949 and 1998. *Science* 292: 2320–2322.

Food and Agriculture Organization of the United Nations (FAO). 2016. *Forestry for a Low-Carbon Future: Integrating Forests and Wood Products in Climate Change Strategies* (FAO Forestry Paper 177). Rome, Italy.

Food and Agriculture Organization of the United Nations (FAO). 2002. *Trees Outside Forests: Towards a Better Awareness* (FAO Conservation Guide No. 35).

Fuss, S., Lamb, W.F., Callaghan, M.W., Hilaire, J. et al. 2018. Negative emissions—part 2: Costs, potentials and side effects. *Environ. Res. Lett.* 13 (6). https://doi.org/10.1088/1748-9326/aabf9f.

Galik, C., Murray, B.C., Mitchell, S., Cottle, P. 2016. Alternative approaches for addressing non-permanence in carbon projects: an application to afforestation and reforestation under the Clean Development Mechanism. *Mitigation and Adaptation Strategies for Global Change* 21: 101–118.

Grassi, G., Stehfest, E. Rogelj, J., van Vuuren, D., Cescatti, A., House, J., Nabuurs, G.-J., Rossi, S. et al. 2021. Critical adjustment of land mitigation pathways for assessing countries' climate progress. *Nature Climate Change* 11: 425–434.

Guo, Z.D., Hu, H.F., Pan, Y.D., Birdsey, R.A., Fang, J.Y. 2014. Increasing biomass carbon stocks in trees outside forests in China over the last three decades. *Biogeosciences* 11: 4115–4122.

Hou, J.Y., Yin, R.S. 2022. How significant a role can China's forest sector play in decarbonizing its economy? *Climate Policy* 23 (2): 226–237. doi:10.1080/14693062.2022.2098229.

Hou, J.Y., Yin, R.S., Wu, W.G. 2019. Intensifying forest management in China: What does it mean, why, and how? *Forest Policy and Economics* 98: 82–89.

International Panel on Climate Change (IPCC). 2006. *Guidelines for National Greenhouse Gas Inventories*. Available at: www.ipcc.ch/report/2006-ipcc-guidelines-for-national-greenhouse-gas-inventories/.

International Panel on Climate Change (IPCC). 2019. *Climate Change and Land: Summary for Policymakers*. Available at www.ipcc.ch/srccl/.

Jiang, F., Chen, J.M., Zhou, L.X., Ju, W.M., Zhang, H.F., Machida, T., Ciais, P., Peters, W., Wang, H.M., Chen, B.Z., Liu, L.X., Zhang, C.H., Matsueda, H., Sawa, Y. 2016. A comprehensive estimate of recent carbon sinks in China using both top-down and bottom-up approaches. *Scientific Reports* 6 (22130).

Jiang, K.J. 2014. Secure low-carbon development in China. *Carbon Management* 3: 333–335.

Johnston, C.T., Radeloff, V.C. 2019. Global mitigation potential of carbon stored in harvested wood products. *PNAS* 116: 14526–14531.

Jonsson, R., Rinaldi, F., Pilli, R., Fiorese, G., Hurmekoski, E., Cazzaniga, N., Robert, N., Camia A. 2021. Boosting the EU forest-based bioeconomy: Market, climate, and employment impacts. *Technological Forecasting & Social Change* 163 (120478).

Ke, S.F., Qiao, D., Yuan, W.T., He, Y.J. 2020. Broadening the scope of forest transition inquiry: What does China's experience suggest? *Forest Policy and Economics* 118 (102240).

Li, P., Zhu, J., Hu, H., Guo, Z., Pan, Y., Birdsey, R., Fang, J.Y. 2016. The relative contributions of forest growth and areal expansion to forest biomass carbon. *Biogeosciences* 13: 375–388.

Liu, P., Yin, R.S. 2012. Sequestering carbon in China's forest ecosystems: Potential and challenges. *Forests* 3: 417–430.

Liu, P., Yin, R.S., Li, H. 2016. China's forest tenure reform and institutional change at a crossroads. *Forest Policy and Economics* 72: 92–98.

Mallapaty, S. 2020. How China could be carbon neutral by mid-century? *Nature* 586: 482–483.

National Forestry and Grassland Administration of China (NFGA). 2019. *China's Forest Resource Reports of the Ninth National Inventories*. Beijing, China.

National Forestry and Grassland Administration of China (NFGA). 2021. *China's Forestry Development Planning for the 14th Five-Year Period (2021–2025)*. Beijing, China.

Ogle, S.M., Kurz, W.A. 2021. Land-based emissions. *Nature Climate Change* 11: 382–383. https://doi.org/10.1038/s41558-021-01040-7.

Piao, S.L., Fang, J.Y., Ciais, P., Peylin, P., Huang, Y., Sitch, S., Wang, T. et al. 2009. The carbon balance of terrestrial ecosystems in China. *Nature* 458: 1009–1013.

Sedjo, R.A., Sohngen, B. 2012. Carbon sequestration in forests and soils. *Annual Review of Resource Economics* 4: 127–144.

Sun, X.F., Canby, K., Liu, L.J. 2016. *China's Logging Ban in Natural Forests: Impacts of Extended Policy at Home and Abroad*. Forest Trends (Information Briefing, March).

Tang, X.L., Zhao, Z., Bai, Y.F., Tang, Z.Y. et al. 2018. Carbon pools in China's terrestrial ecosystems: New estimates based on an intensive field survey. *PNAS* 115 (16): 4021–4026.

United Nations Framework Convention on Climate Change (UNFCCC) Secretariat. 2020. *Reference Manual for the Enhanced Transparency Framework under the Paris Agreement.* Bonn, Germany.

Wang, Y.F., Li, L.Y., Yin, R.S. 2021. A primer on forest carbon policy and economics under the Paris Agreement. *Forest Policy and Economics* 132 (102595).

Xinhua. 2020. *Remarks by Chinese President Xi Jinping at the Climate Ambition Summit.* Available at: www.xinhuanet.com/english/2020-12/12/c_139584803.htm.

Yin, R.S. 2009. *An Integrated Assessment of China's Ecological Restoration Programs.* Dordrecht, The Netherlands: Springer.

Yin, R.S., Sedjo, R.A., Liu, P. 2010. The potential and challenges of sequestering carbon and generating other services in China's forest ecosystems. *Environmental Science & Technology* 44: 5687–5688.

Zeng, W.S., Tomppo, E., Healey, S.P., Gadow, K.V. 2015. The National Forest Inventory in China: History, results, and international context. *Forest Ecosystems* 2. Available at: https://doi.org/10.1186/s40663-015-0047-2.

Zhou, S., Wang, Y., Yuan, Z., Ou, X. 2018. Peak energy consumption and CO2 emissions in China's industrial sector. *Energy Strategy Review* 20: 113–120.

## Appendix I. Quantifying carbon emissions and removals

### *Projecting aggregate emissions*

The GCAM-TU is a modified version of the GCAM developed by the US Pacific Northwest National Lab, which solves intertemporally for the optimal equilibrium prices and quantities of energy goods, emissions, and agricultural products. To better reflect China's reality, the GCAM-TU has adjusted the economic and technological structures and updated certain assumptions of the GCAM accordingly (Duan et al. 2021). It has been used for exploring China's carbon peaking and GHG mitigation roadmaps (Zhou et al. 2018). The IPAC model was constructed by the Energy Research Institute of the National Development and Reform Commission of China, and it has been used to design China's emissions scenarios and global emissions scenarios (Jiang 2014). Built by Italian scholars, the WITCH model takes into account innovations in energy efficiency and clean technologies resulting from research and development activities (Bosetti et al. 2006), and it has been employed in global multi-model comparisons of emission reduction pathways and climate policies (Aldy et al. 2016). A detailed overview of these and other models can be found in the Supplementary Materials of Duan et al. (2021), particularly Table S1.

Based on limiting the carbon emissions to a temperature rise within 1.5°C, the above models took 2005 as the initial year of operation, with 2005–2015 being the calibration period before making projections in every five years. The WITCH model was run until 2060, whereas the GCAM-TU and IPAC model made projections up to 2050. In the case of WITCH model projections, the outcomes for the intermittent years were derived by interpolating the original projections. With GCAM-TU and IPAC model projections, I assumed zero emissions in 2060 and interpolated the outcomes for the remaining years. The very low emission levels by 2050 should alleviate any concern about the credibility of this assumption.

By comparison, the WITCH model requires a higher-level carbon capture and storage, in addition to a more drastic early emission reduction (Bosetti et al. 2006), while the GCAM-TU and IPAC projections have thus far come closer to what has occurred. The GCAM-TU and IPAC projections prescribe possibly more realistic paths for China to reach its targets of emission reduction and removal enhancement, as well as those of green investment, economic growth, and technical change (Zhou et al. 2018, Jiang 2014). Thus, these two sets of projections—GCAM-TU and IPAC—were used in my analysis.

*Forecasting ecosystem sequestration*

The targets of the National Forest and Grassland Administration are to raise the country's forest cover to 25% of the land surface, or $239.98 \times 10^6$ha, by 2030, and to 26%, or $249.58 \times 10^6$ha, by 2035 (NFGA 2019). Knowing that it would be more feasible to add almost $10 \times 10^6$ha of new forests within each of the two remaining decades, I chose to let the area reach $249.58 \times 10^6$ha by 2040; afterward, it is assumed to stay at that level until 2060. The unit stocking level from 2019–2020 was assumed to maintain the same growth rate as that during the 9$^{th}$ NFI. Later, an annual growth rate of 0.67% is maintained until 2035, which was prorated on the basis of the NFGA target for the stock growth of arbor forests (from 95m$^3$/ha to 105m$^3$/ha) by 2035 (Piao et al. 2009). Then, an annual growth rate of 0.89% until 2050 was adopted, which was also prorated based on the NFGA target for arbor forest (from 105m$^3$/ha to 120m$^3$/ha) by 2050 (NFGA 2019). The same rate of annual growth was assumed to continue for the last decade (2051–2060). We call this a conventional case.

The biomass carbon stock is a product of the aggregate stock volume multiplied by the biomass expansion factor (1.045798) and the carbon conversion factor (0.48899) (Fang et al. 2001, Tang et al. 2018), which has been documented in the latest government report (NFGA 2019). Litter and deadwood are included in the biomass pool. For making a comparison and exploring potential, I also constructed a case of accelerated forest growth, which is if more substantive and progressive silvicultural practices, including thinning and other stand treatments, could be implemented (Hou et al. 2019, Liu and Yin 2012). Under this case, the annual growth rate would be 0.85% from 2021 to 2030, 0.90% from 2031 to 2040, and 0.95% from 2041 to 2060. Considering the international experience and the status of China's forests—a dominance of premature stands with low stocking levels (Ke et al. 2020, Hou et al. 2019)—I argue that this is a sensible case that the country should strive for.

The soil carbon fluxes have not been continuously inventoried because of the efforts and costs involved (Piao et al. 2009), and the only estimate of interest came from Jiang et al. (2016), which derived soil C fluxes through modeling. According to that study, China's forest biomass carbon pool had a gain of 0.13 ± 0.04Pg per year during the 2000s, while the soil carbon pool accrued 0.064 ± 0.03Pg per year in the same period. To be realistic, I took their mean values, resulting in 67% of the sequestered carbon distributed in biomass and 33% in soil. Again, the dominance of young stands of trees established recently on marginal lands or restored on degraded natural forests suggests that it is unlikely for the forests to accumulate a higher percentage of soil organic matter in the near term (Piao et al. 2009).

### *Estimating HWP storage*

Johnston and Radeloff (2019) first calibrated the Global Forest Products Model with the *FAOSTAT* data up to 2015 before projecting the productions, imports, and exports of different products. Their projections were based on shared socioeconomic pathways (SSP) describing different futures of multiple socioecological factors, with the cross-border leakages, induced by trade flows, being incorporated. From the projected outcomes, the authors further inferred global consumptions as well as individual countries' consumptions. Also, the country-specific forest area changes over time under different SSPs were defined as a function of the corresponding demographic and economic dynamics and represented through an Environmental-Kuznets-Curve relationship (Choumert et al. 2013). Then, the changes in forest stock volume were derived.

I adopted their SSP4 and SSP5—the two extreme cases—to reflect the probable range of C stored in HWPs, and the projected storages in every five years were annualized. More details on the key components of these and other SSPs, and the main elements of the forest sector pathways corresponding to each SSP can be found, respectively, in Table S3 and Table S4 of the article.

# 6 Carbon leakage induced by forest products trade

Outcomes and implications of alternative accounting practices for China

**Introduction**

International trade separates the consumption of goods from the environmental effects that are generated in their production (Rodrik 2022, Branger and Quirion 2014, Antweiler et al. 2001). This, of course, includes the trade of farm and forest products, whereby greenhouse gases (GHGs) are emitted from agriculture, forestry, and other land uses (AFOLU), as well as the changes of these land uses over time (Hong et al. 2022). As the global community gears up to implement the Paris Agreement on combating climate change, it is relevant and timely to investigate issues surrounding GHG emissions embodied in the trade of forest products (Böhringer et al. 2022, Yin 2021), which is what I do in this chapter.

To my knowledge, the two intertwined issues that have attracted broad attention are carbon leakage (Hong et al. 2022, Hou et al. 2022) and border tax adjustment (Böhringer et al. 2022, Frankel 2022). Carbon leakage results from the difference between the production emissions of a country and its consumption emissions (Hong et al. 2022, Babiker 2005). Over the past several decades, as economic growth and globalization continued, the international trade of finished goods, as well as intermediate goods and materials, increased tremendously, resulting in greater flows of embodied GHGs (Hong et al. 2021, Branger and Quirion 2014, Antweiler et al. 2001). Cross-country carbon flows can be measured from the perspective of consumption or production (Yan et al. 2013, Babiker 2005). Different perspectives, however, may lead to different emission outcomes and thus mitigation responsibilities (Branger and Quirion 2014, Ritchie 2019).

When countries set targets or make international comparisons of emission reductions, they tend to look at production-based emissions—$CO_2$ and other GHGs emitted within a producer country's own borders. In fact, the production-based approach is still used in the rulings of trade disputes by the World Trade Organization (WTO) and the national emissions inventory under the current Intergovernmental Panel on Climate Change guidelines (IPCC 2006, Hong et al. 2022). Nevertheless, this approach fails to capture the net emissions embodied in trade—emissions derived from imported goods minus the emissions derived

DOI: 10.4324/9781003436652-7

from exported goods (Ritchie 2019, Babiker 2005). Consequently, the mitigation responsibilities of different countries can be distorted.

As an example, *Our World in Data* [1] reports that China's net export of $CO_2$ in 2020 was equivalent to 8.42% of its domestic emissions; that is, its consumption-based emissions were 8.42% lower than its production-based emissions. On the other hand, the United States' net import of $CO_2$ in 2020 was equivalent to 10.21% of its domestic emissions, implying that the US consumption-based emissions were 10.21% higher than its production-based emissions. *Our World in Data* also estimates that emissions produced by China in 1990 were already 6% higher than those measured from the consumption perspective. As shown in Figure 6.1, this gap enlarged to 22% in 2006 before declining to less than 10% more recently. In other words, the difference between China's consumption- and production-based accounting for $CO_2$ reached 1.41 gigatons (Pg) in 2006, and it remained at 0.93Pg in 2020.

As a result, there have been increasing calls for adopting a consumption-based approach to international carbon accounting (e.g., Peters and Hertwich 2008, Yan et al. 2013). It is argued that establishing a GHG emission catalogue based on consumer responsibility would not only help the international efforts of emissions reduction but also facilitate the formation of comparative advantages of low-carbon products and the transfer of clean technologies (Rodrik 2022, Nordhaus

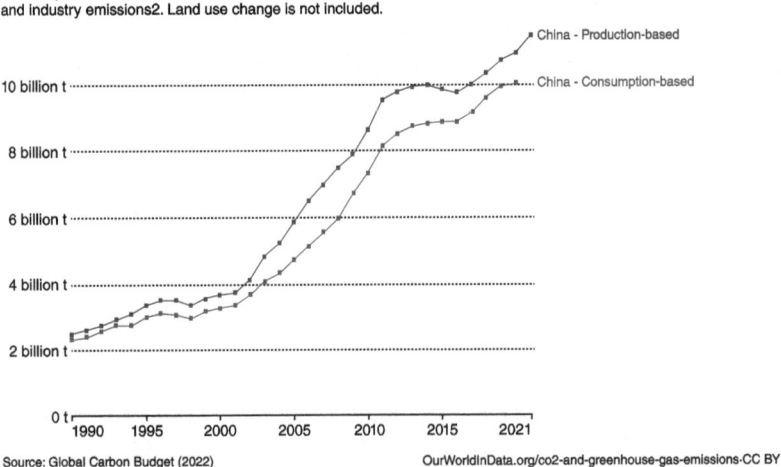

Consumption-based emission[1] are national emissions that have been adjusted for trade. This measures fossil fuel and industry emissions[2]. Land use change is not included.

Source: Global Carbon Budget (2022)  OurWorldInData.org/co2-and-greenhouse-gas-emissions-CC BY

**1. Consumption-based emissions:** Consumption-based emissions are national or regional emissions that have been adjusted for trade. They are calculated as domestic (or 'production-based' emissions) emissions minus the emissions generated in the production of goods and services that are exported to other countries or regions, plus emissions from the production of goods and services that are imported. Consumption-based emissions = Production-based – Exported + Imported emissions

**2. Fossil emissions:** Fossil emissions measure the quantity of carbon dioxide (CO2) emitted from the burning of fossil fuels, and directly from industrial processes such as cement and steel production. Fossil $CO_2$ includes emissions from coal, oil, gas, flaring, cement, steel, and other industrial processes. Fossil emissions do not include land use change, deforestation, soils, or vegetation.

*Figure 6.1* China's consumption- vs. production-based $CO_2$ emissions
Source: *Our World in Data*

2021). Thus, it is both economically sound and morally sensible for the global community, including Parties to the United Nations Framework Convention on Climate Change (UNFCCC), to confront carbon leakage induced by commodity trade (Pan et al. 2008, Ghertner and Fripp 2007).

Moreover, carbon leakage raises concern about the effectiveness of unilateral climate policies (Böhringer et al. 2022, Frankel 2022). In general, economists worry that leakage hampers the global cost-effectiveness of emission reduction. Representatives of emissions-intensive and trade-exposed (EITE) industries, such as iron and steel, cement, fertilizers, aluminum, and electricity generation, in climate-concerned nations point out that leakage results from an undue disadvantage vis-à-vis less regulated competitors abroad. To deal with this problem, some kind of broader adjustment mechanism (BAM) can be put in place by applying the domestic carbon price to emissions embodied in traded goods, which is thought to 'level the playing field' for EITE industries and thus avoid carbon leakage through the competitiveness channel (Frankel 2022), as elaborated later.

Box 6.1 below details the announcement made by the European Commission (EC) to impose a carbon BAM—an import tax designed to corral other countries into tackling climate change. Despite its appeal to the EC, however, this measure, coupled with the current production-based international carbon accounting, constitutes a 'double whammy' to the exporting countries, most of which are developing and emerging economies (Branger and Quirion 2014, Ghertner and Fripp 2007). That is, the BAM, if implemented, could cause detrimental effects on not only global trade itself but also international efforts of climate change mitigation. Thus, the BAM seems inconsistent with the general principle of "common but differentiated responsibilities and respective capabilities," adopted by the UNFCCC in 1992.

---

**Box 6.1 European Union's carbon border tax—An excerpt from Böhringer et al. (2022)**

Reaching the EU's goal of cutting net greenhouse-gas emissions by 55% by 2030 will require it to reduce emissions both at home and beyond its borders. To this end, the European Commission announced in July 2021 that it would include a carbon border adjustment mechanism (BAM)—an import tax—in its initiative aimed at meeting its emissions-reduction target. The EU would thus tax imported goods sold in its markets based on their carbon contents. The proposed levy is intended to address so-called carbon leakage, which occurs when businesses in the EU move production to non-member countries with less stringent emissions rules.

The BAM will initially apply to the emissions-intensive and trade-exposed industries most at risk of leakage and will likely be expanded to other sectors in the coming years. Currently, EU-made products in these industries are covered under domestic carbon pricing, but those from outside the bloc are not. If a country already has a domestic carbon price and/or tax, the border tax will be lowered or waived to encourage countries to price carbon in their own markets. Those that cannot or will not institute a carbon pricing will have to pay the full levy. The EU tax will be phased in over the next a few years. By 2023, importers will be required to report emissions embedded in the goods they import, though the tax on those emissions will not be imposed until 2026. The €1 billion of annual revenue expected from the BAM, as well as the €9 billion in annual revenue expected from the EU Emissions Trading System from 2023–2030 and taxes on multinational corporations, will support the Union's €750 billion COVID-19 pandemic recovery fund.

Though not yet approved, the proposed tax is already influencing the decisions of policymakers and companies in the EU's trading partners. The trade protectionism that it entails risks hurting developing countries. For example, Turkey and Indonesia plan to introduce carbon taxes to mitigate the BAM's effects on their economies. Rather than addressing only carbon leakage and leaving developing countries to adapt as best they can, the EU should allocate part of the revenue from the proposed BAM to help foster a just green transition for poorer countries.

This chapter examines the intertwined issues of carbon leakage and border adjustment by assessing the outcomes of alternative accounting approaches and analyzing the pros and cons of border tax adjustment from different perspectives. The following section synthesizes the primary findings of two case studies as well as the data and methods they used in determining carbon leakage, found in the latest literature. We will then examine the pros and cons of border tariff adjustment. The next section assesses the potential carbon leakage induced by China's trade of forest products—solid wood, panel, and paper and paperboard products. It relies on China's experience once more, due to that country's significance in international climate mitigation, its large role in world trade, and also the availability of its data. Finally, discussion and closing remarks are given in the final section. I hope that this chapter will not only bring attention to the relevant issues in relation to forest products trade but also underscore the implications to forest sector climate solutions. In preparing this chapter, I have taken the liberty of drawing information from Hong et al. (2022), Hou et al. (2022), Böhringer et al. (2022), and Rodrik (2022), among others.

**Case studies of carbon leakage embodied in trade**

The task of this section is to synthesize the main results as well as the analytic procedures and data used, in two of the most recent studies—Hong et al. (2022) on land-use emissions embodied in international trade and Hou et al. (2022) on China's trade of paper and paperboard products (heretofore, paper products).

## Global land-use emissions

Global analysis of trade-related land-use emissions has been hampered by a lack of detailed estimates of country-, region-, and product-specific land-use emissions. Thanks to the notable efforts of Hong et al. (2021), we now know that land-use emissions represent about 25% of net global anthropogenic GHG emissions in recent years, which is substantial enough to threaten international climate goals even if fossil fuel emissions are drastically reduced. Hong et al. (2022) attributed land-use emissions associated with the production of various farm and forest products to final consumption in 141 world regions (mostly individual countries), using a multiregional input-output (MRIO) model and an integrated dataset for the years of 2004, 2007, 2011, 2014, and 2017[2]. The input-output approach maps emissions to the region where related goods are ultimately consumed, even if the products are transshipped through an intermediary region or are intermediate constituents in a multiregional supply chain. The difference between each region's production emissions and consumption emissions defines the net emissions embodied in trade.

It was found that during 2004–2017, roughly one billion hectares of farmland and pastureland were used for traded agricultural products, representing ~22% of agricultural land worldwide. Meanwhile, a higher share of global land-use emissions (~27%) were embodied in international trade, ranging between 4.5 and 5.8Gt $CO_2$e per year. Moreover, 16.7% of the global land-use emissions in 2017 came from the trade of wood products, or 4.1% of the total global anthropogenic GHG emissions (Hong et al. 2021). Overall, a dominant global feature of embodied land-use emissions, persistent over time, was large exports of emissions from countries such as Brazil, Indonesia, Argentina, Australia, and Canada to consumers in developed regions such as the US, Europe, and Japan. Put differently, consumption-based land-use emissions were considerably lower than production-based emissions in Oceania, Southeast Asia, and Latin America, whereas the opposite is true in Europe, North America, and East Asia. Furthermore, China had become an important importer of agricultural and forest products by 2017 as well as the largest net importer of embodied land use emissions. The corresponding amount of land-use emissions was 814 million tons (Mt) of $CO_2$e, followed by the US, Japan, Germany, the UK, Italy, South Korea, and Saudi Arabia (each in the range of 100 to 370Mt $CO_2$e).

## Emissions of China's paper net exports

China is one of the largest exporters of paper products that are manufactured mostly from wood fibers, based on intense energy use (Hou et al. 2022). Along the deeply connected global value chains (GVCs), the division of labor and specialization of manufacturing activities have enabled certain countries, mostly advanced economies, to shift the production of some consumer goods to other countries, mainly developing ones (Pietrobelli and Staritz 2018). The production-based method of carbon accounting, however, neglects the impact of the specialization and segmentation of GVCs on $CO_2$ emissions, resulting in potentially large leakages embodied in trade flows. Hou et al. (2022) further exposited that in addition to the imperative to quantify the carbon flows along the GVCs, another driver for China to consider consumption-based carbon accounting would indeed be the fact that the EU announced the adoption of a BAM by applying the domestic carbon pricing to emissions embodied in imported goods. This policy has made it not just necessary but also urgent to examine $CO_2$ flows across borders and determine the responsibility of emission reduction more fairly.

Using an MRIO model and statistics from the World Input-Output Database, similar to Hong et al. (2022), Hou et al. (2022) traced carbon emissions embodied in China's paper trade from 2000 to 2014. By dividing the traded paper products into intermediate products and final consumer products when calculating the embodied carbon of a given country, they were able to show that China's paper industry relied heavily on the import of raw materials to export a large quantity of finished paper products. The amount of $CO_2$ emissions of China's paper industry was estimated to be about 64Mt in 2000, but it rose to 152Mt in 2014.

Moreover, Hou et al. (2022) revealed that carbon emissions embodied in China's paper exports were much larger than those of its imports. On the one hand, annual $CO_2$ emission embodied in imported finished as well as intermediate paper products rarely went above and beyond 10Mt during 2000–2014. On the other, the amount of annual $CO_2$ emissions embodied in exported finished as well as intermediate paper products were around 33Mt in 2000, and it rose to roughly 83Mt in 2008 before declining to 47Mt by 2014. As a result, China's exports of large quantities of paper products led to high carbon emissions within China. However, its emissions were much lower if determined from a consumption perspective. In 2014, for example, the amount of $CO_2$ emissions embodied in net export was 38.33Mt (see Table 6.1). In comparison, the US had the largest consumption-based emissions among all the major trading partners of China. Its amount of $CO_2$ emissions embodied in net import was 12.80Mt.

In short, Hong et al. (2022) and Hou et al. (2022) have convincingly demonstrated the need for a consumption-based approach to carbon accounting to identify GHG emissions embodied in trade flows. It is thus worthwhile to conduct the trade-induced accounting of global land-use emissions and determine their drivers to better target and coordinate international mitigation efforts (Hong et al. 2022). Similarly, the study of net exports of China's paper industry and its induced leakage of $CO_2$ emissions by Hou et al. (2022) has provided an interesting example for identifying the direction and magnitude of embodied carbon flows in the broader context of forest products trade—the trade of not just paper products but also solid wood and panel products in China or elsewhere.

*Table 6.1* Carbon emissions embodied in China's paper trade in 2014

| Country | Production-based | Consumption-based |
|---|---|---|
| Austria | 1,753.33 | 208.27 |
| Belgium | 623.43 | 462.60 |
| Canada | 4,444.28 | 1,695.97 |
| Czech Republic | 471.94 | 377.48 |
| Finland | 2,985.68 | 291.03 |
| France | 2,954.32 | 1,353.68 |
| Germany | 7,617.99 | 3,898.47 |
| Italy | 5,007.80 | 2,692.01 |
| Japan | 11,595.23 | 2,372.80 |
| Korea | 2,494.18 | -423.82 |
| Netherlands | 859.70 | 717.11 |
| Poland | 2,380.18 | 1,572.11 |
| Portugal | 1,145.10 | 495.72 |
| Spain | 3,897.16 | 1,438.07 |
| Sweden | 808.20 | 388.65 |
| Switzerland | 316.50 | 128.51 |
| The United Kingdom | 3,588.15 | 1,983.12 |
| United States | 35,396.14 | 22,595.12 |
| Brazil | 4,980.85 | 2,261.11 |
| China | 46,591.53 | 8,262.37 |
| Indonesia | 5,204.29 | 598.19 |
| Russia | 2,213.99 | 3,948.27 |

*Note*: Results reported here were derived from Figure 3 of Hou et al. (2022)

For one thing, the whole forest products industry of China is much larger than its paper industry—the annual gross manufacturing value of China's forest products industry reached 2.34 trillion Yuan in 2017 (Ke et al. 2020). Of that total, 1.24 trillion Yuan came from the processing of solid wood products, including furniture, with the balance being panel and paper products. Certainly, this makes the task of quantifying the carbon flows embodied in the forest products trade a more important and relevant research task. But Hou et al., or someone else, was unable to complete it earlier. At my request, they have finally obtained some preliminary results, which will be presented in a later section of this chapter.

## Boarder carbon adjustments

Böhringer et al. (2022) reviewed the potential impacts of border carbon adjustments on leakage reduction, competitiveness restoration, cost-effectiveness, and equity and cooperation enhancement. They argued that the fundamental challenge in climate policy is to cope with the fact that climate protection constitutes a global public good. Each country has a strong incentive to a free ride on the emissions abatement of other countries while contributing little itself. Without a supranational authority to coerce common action, a 'tragedy of the commons' arises, in which countries' self-interest prevents a globally efficient climate policy. Because of free-rider incentives, the voluntary commitments made by countries are not sufficient to achieve the Paris Agreement target of limiting global warming to well below 2°C. Therefore, cost-effective climate action calls for coordinated, harmonized carbon pricing across domestic and international jurisdictions (Stern 2006, Mehling et al. 2018).

More specifically, when carbon emissions are priced unilaterally, the global environmental impact will be undermined to the extent that international markets transmit spillover effects that increase emissions in other countries, leading to carbon leakage (Clausing and Wolfram 2023). These effects occur primarily through two channels—competitiveness and technological transfer. Carbon leakage through the competitiveness channel occurs when unilaterally regulated carbon-intensive businesses reduce production because of higher operating costs, while production by less regulated manufactures abroad increases. The other potential leakage channel, international technology spillover, is also possible but more uncertain and thus often ignored in quantitative analyses. Environmentalists view leakage as an avoidance mechanism—a loophole whereby consumers in developed countries substitute higher priced domestic products of lower carbon content with imports that are cheaper but more carbon intensive. As highlighted by various studies, again, China is by far the most important target for import control by the EU and some other trading partners (e.g., Hong et al 2022, Rodrik 2022).

To address these concerns, countries with a domestic carbon price could implement a BAM, that is, levy a charge on the carbon embodied in imports from regions without (adequate) carbon pricing (Clausing and Wolfram 2023). For the sake of symmetry, a rebate to exports (to unregulated regions) could offset their embodied carbon payments. The adjustment has the effect of taxing consumption rather than production. By capturing emissions embodied in traded goods, a BAM can reduce leakage through the competitiveness channel and improve the global cost-effectiveness of unilateral action—the global economic cost to achieve a given reduction in global emissions. Import adjustments also mean that suppliers from countries without domestic carbon pricing face an increase in carbon costs so that they do not gain an unfair competitive advantage when carbon prices rise. Similarly, export rebates create competitive neutrality in foreign markets. Tariffs not just level the playing field by making it harder for industries to offshore their emissions to countries with lower regulatory standards and compliance costs, but also encourage governments to join the club of countries making serious commitments and then to abide by them (Frankel 2022). In general, BAMs can secure competitive neutrality across countries with differential carbon prices if foreign countries with some carbon pricing also adopt BAM or if the domestic unilateral adjustment mechanism covers only the difference in carbon prices.

In this case, differences in carbon prices between countries don't incentivize production shifts to the country with the lower carbon price. However, the carbon content of goods produced in foreign jurisdictions is difficult to measure, and empirical evidence on leakage remains scarce, as shown above. More specific targeting and legal verification of emissions streams would lead to higher implementation costs and technical barriers to trade. Sector-specific simulation studies found leakage rates that range from 20% to 70% for the EITE industries, suggesting that a significant portion of emissions reductions in domestic production could be offset by increased emissions abroad. Also, the viability of implementing a BAM can be substantially reduced with the current legal and practical constraints in the EU or elsewhere (Böhringer et al. 2022).

It can be further argued that a BAM, if adopted in combination with a production-based carbon accounting, will hurt the exporting counties unfairly (Durántez and Obern 2022), and it is not in line with the UNFCCC principle of "common but differentiated responsibilities and respective capabilities." Therefore, if a BAM is implemented, it should be implemented in conjunction with a consumption-based international carbon accounting, which is ethically more justifiable. The legality of a BAM may have to be adjudicated by the WTO under the current regime of international trade. EC officials claim that its proposed BAM will comply with WTO rules, and conventional wisdom suggests that WTO members are already permitted to enact trade barriers for environmental ends, so long as the measures do not unfairly discriminate against foreign firms (Frankel 2022). So, it seems that adjudicating $CO_2$ border taxation would be a natural job for a revived WTO, if its members were to give it the mandate. For instance, it could take a case to the WTO if the EU BAM were to become protectionist, whether by including too many industries with indirect and hard-to-quantify $CO_2$ use or by overestimating the gap in the effective prices of carbon in the EU.

Nonetheless, the WTO rules are generally predicated on where a good is produced rather than where it is consumed (Frankel 2022). Without harmonizing the fundamental principles of international trade and climate policy, it is likely that the WTO ruling would hurt the exporting economies (Rodrik 2022). Given that humanity is facing an existential threat of climate change, the WTO, IPCC, and UNFCCC should consider modifying, updating, and harmonizing their rules and regulations. To enable the net-zero transition and foster a dynamic and inclusive economy, therefore, competition authorities, including the WTO, "must recognize the scale of the task they face and shake off outdated modes of thinking—the sooner, the better" (Coyle 2022). More broadly, the climate challenge is too important to be overshadowed by trade and competitiveness consideration (Rodrik 2022). Thus, the WTO may not be the right forum for determining the circumstances under which trade rules

106  *Carbon leakage induced by forest products trade*

should constrain climate policy and green development measures. It would be difficult, and perhaps inappropriate, for global trade panels to render judgement on the availability of alternative instruments and second-guess domestic political tradeoffs.

## Leakage from China's forest products trade

China's forest products industry's basic feature is to import raw material (roundwood) and preliminarily processed forest products (pulp, recycled papers, and lumber) and to manufacture and re-export finished forest products (furniture, panel, and paper products) (Ke et al. 2020). As shown in Figure 6.2, China's forest products industry has been exporting large quantities of finished products, especially panel and furniture products, using imported as well as domestic raw materials and intermediate products, such as roundwood, lumber, and pulp. This suggests a relatively high amount of $CO_2$ emissions if counted from the perspective of consumption instead of that of production. However, the calculation must incorporate energy use in addition to the use of raw material and intermediate inputs, no matter if domestically produced or imported.

With my encouragement and at my request, Professor Fangmiao Hou, the lead author of Hou et al. (2022), has replicated their paper industry study for the whole forest products industry of China. Interested readers can refer to that article for the basic methods used in deriving these results. The estimated $CO_2$ emission coefficient of the forest products industry shows a downward trend from 2000 to 2015 (Figure 6.3), largely

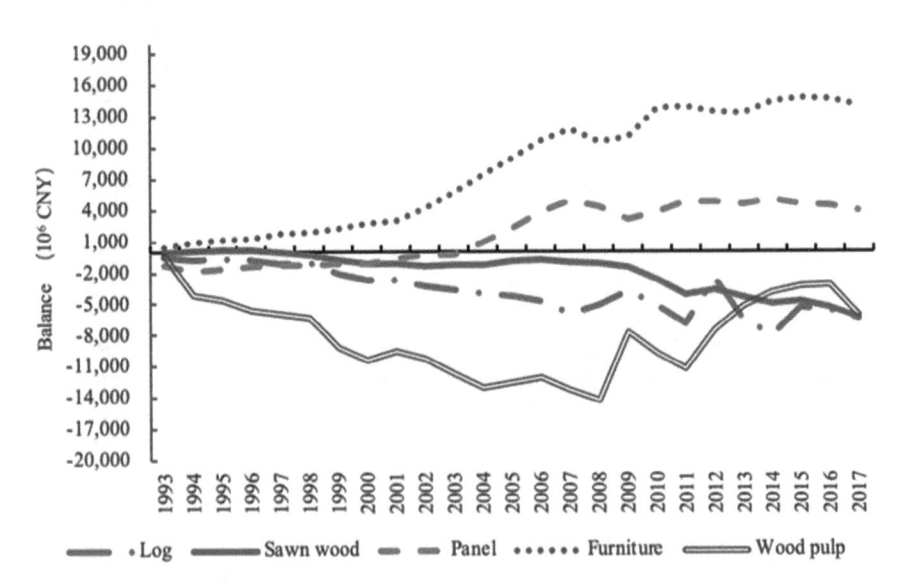

*Figure 6.2* China's trade balance of major forest products (in 2017 constant price)
Source: Ke et al. (2020)

```
9
8
7
6
5
4
3
2
1
0
   2000 2001 2002 2003 2004 2005 2006 2007 2008 2009 2010 2011 2012 2013 2014 2015
```
— direct carbon emission coefficient
— complete carbon emission coefficient
— ratio of complete carbon emission coefficient to direct carbon emission coefficient

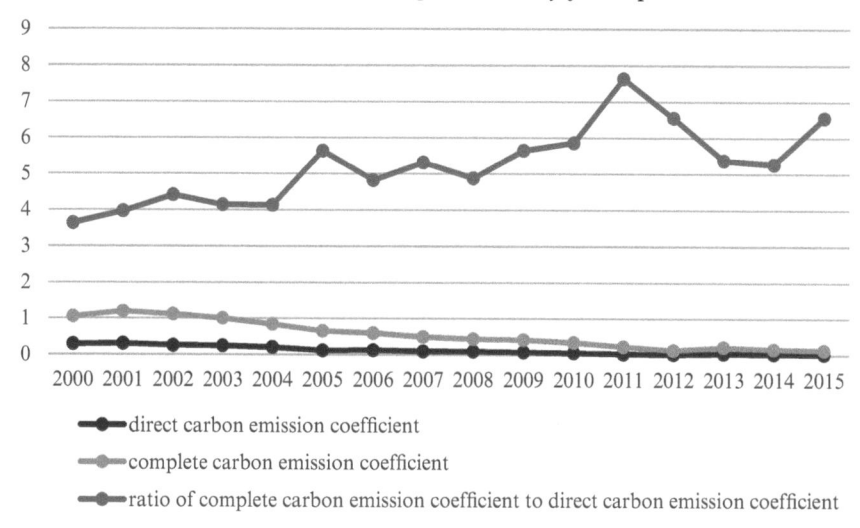

*Figure 6.3* Estimated coefficients of $CO_2$ emissions for China's forest products industry from 2000 to 2015

due to the elimination of low productivity in the manufacturing process, improvement of material utilization, and increased use of cleaner energy. The total emission coefficient is greater than the direct emission coefficient, indicating that the forest products industry, while using its own materials, gradually became more reliant on other energy sources (e.g., electricity), causing an increase in the embodied $CO_2$ emissions in other industries.

From 2000 to 2005, China's $CO_2$ emissions embodied in exported forest products showed a trend of rapid growth, which had to do with China's accession to the WTO in 2001 and its concomitant expansion of international trade. After 2005, the emissions stabilized. But from 2010 to 2014, the industry's export-induced emissions gradually declined, as it promoted the integration of forest resource management and forest products manufacturing, along with improvement of manufacturing efficiency (Jing et al. 2019). Also, many manufacturing patents and process innovations emerged during this period (Tong and Chen 2015), resulting in persistently decreased energy consumption and $CO_2$ emissions. In addition, due to the adverse impact of the 2008 global financial crisis, the export volume in 2009 decreased significantly, causing a corresponding drop in $CO_2$ emissions thereafter (see Figure 6.4). The net emissions between exports and imports were 184.55 Mt of $CO_2$ in 2000, declined to 143.75 Mt by 2010, and then to 114.51 Mt in 2014. Nonetheless, the production-based $CO_2$ emissions of China's forest products industry remained far larger than the consumption-based emissions.

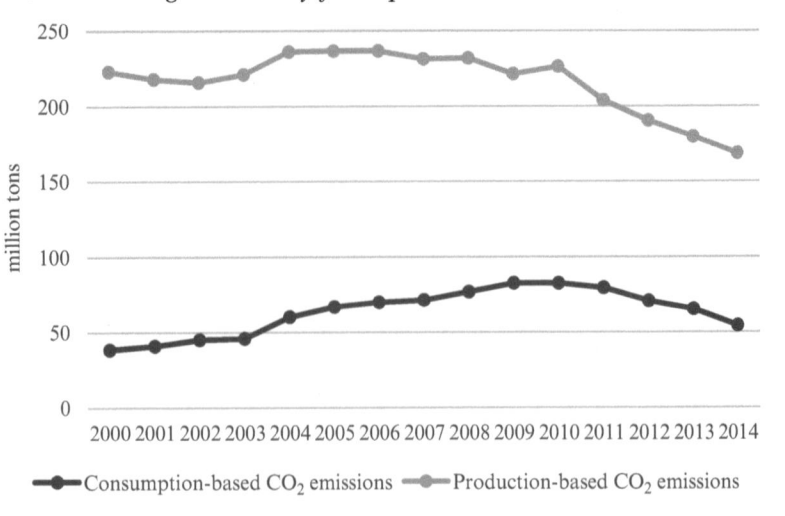

108  Carbon leakage induced by forest products trade

*Figure 6.4* Different measurements of CO$_2$ emissions for China's forest products industry from 2000 to 2014

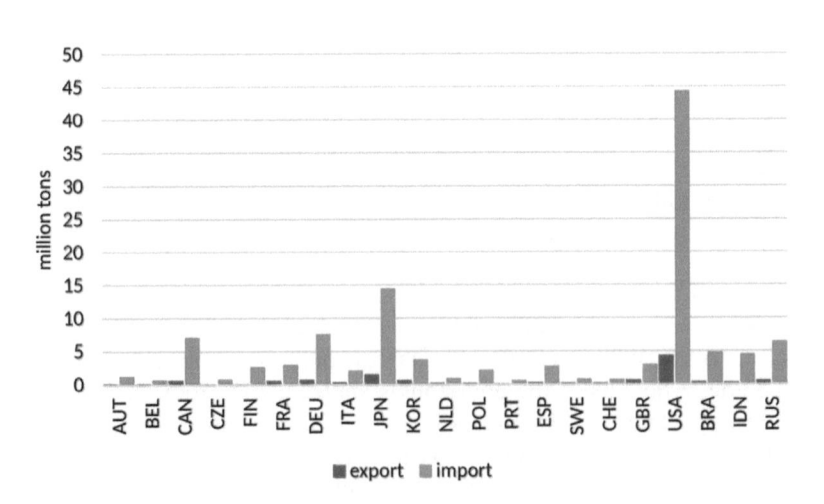

*Figure 6.5* CO$_2$ emissions with 21 trading partners of forest products in 2014
*Note*: The 21 countries from left to right are Austria, Belgium, Canada, Czech Republic, Finland, France, Germany, Italy, Japan, South Korea, Netherlands, Poland, Portugal, Spain, Sweden, Switzerland, United Kingdom, United States, Brazil, Indonesia, and Russian Federation.

A further calculation of the embodied emissions for the 21 major trading partners in 2014 shows that China had the highest amount of CO$_2$ emissions embodied in its exporting forest products—1.3 times that of the United States (Figure 6.5). Then came the United States and Japan, with no other country having a leaked amount of above 10Mt of $CO_2$. From the consumption perspective, however, the $CO_2$ emissions of the United States ranked first—2.7 times that of China, which ranked second, followed by Russia.

Based on the decomposition of the embodied carbon flows of China's forest products industry in 2000 and 2014, it is found that the top three countries of $CO_2$ emissions are the United States, Japan, and Korea (Figure 6.6). In 2000, the embodied $CO_2$ emissions to the United States were 71.12Mt, accounting for 36.5% of the total. Japan was the second, accounting for 21.5%, while South Korea accounted for 4.0%. In 2014, the emissions embodied in its exports to the United States were 204.63Mt, accounting for 31.0% of the total. Japan and Korea accounted for 17.3% and 8.8%, respectively.

## Discussion and closing remarks

It is more appropriate to estimate consumption-based $CO_2$ emissions by correcting trade-induced leakage. Trade-related accounting of global land-use emissions is needed to uncover the international drivers of land-use emissions and to better target and coordinate mitigation efforts (Hong et al. 2022). If it were possible to produce the same crops by sustainable intensification, land-use emissions embodied in trade would be drastically reduced. Likewise, the Hong et al. study makes the case that policies incorporating environmental externalities could help broaden the notion of comparative advantage as well as reduce incentives for importers to offshore GHG emissions and for exporters to backslide into environmentally destructive practices. Mitigation of global land-use emissions and sustainable development will thus depend on improving the transparency of supply chains (IPCC 2020).

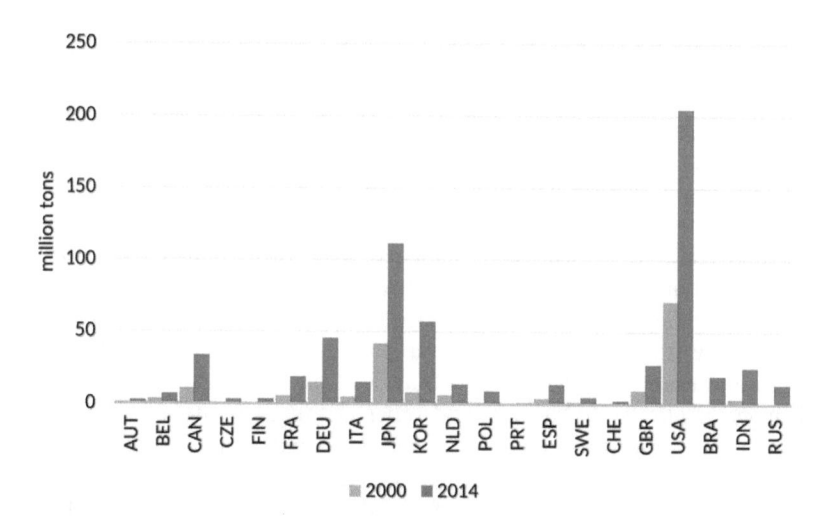

*Figure 6.6* $CO_2$ emissions embodied in China's forest products trade in 2000 and 2014 (see country definitions beneath Figure 6.5)

$CO_2$ emissions embodied in forest products trade is also important, as demonstrated by China's experience. Although the gap narrowed over time, the production-based $CO_2$ emissions of China's forest products industry remained far larger than the consumption-based emissions—114.51Mt in 2014. That amount of leaked emission was over one tenth of China's total $CO_2$ leakage induced by international trade for the same year. A decomposition of the embodied carbon flows of China's forest products industry in 2000 and 2014 further highlighted that its top three destinations of $CO_2$ emission leakage were the United States, Japan, and Korea. Certainly, this kind of analysis is very beneficial to identifying the direction, magnitude, and destination of the emissions embodied in the trade flows of forest products (Hou et al. 2022).

Further, it can be anticipated that more and more national governments and other domestic and international organizations will look for similar assessments of the causes and effects of inter-regional and international GHG emissions induced by land use and land-use change and flows of traded goods. Doing so is an essential component of carbon accounting and mitigation efforts. In this regard, my assessment of the $CO_2$ emissions embodied in China's forest products trade, coupled with the reviewed studies of Hong et al. (2022) and Hou et al. (2022)—the two latest cases based on similar analytic methods and data uses—has opened an important direction for future research.

Nevertheless, consumption-based accounting of land-use emissions or traded goods is not yet required by any national policies or international agreements, and it is unrealistic to expect that carbon leakage in international accounting or border adjustment can be resolved easily or soon. The science and policy community ought to accumulate strong evidence and explore the implications of alternative accounting or adjusting approaches (Rodrik 2022, Frankel 2022). Well-crafted policies will help ensure that any additional costs associated with avoiding such land-use change and trade practice would at least initially be borne by (typically more affluent) importing regions.

Indeed, harmonized carbon pricing across borders is hard to achieve in the real world, even though carbon leakage can reduce the cost-effectiveness of unilateral approaches to abate global emissions. But the EU and other advanced economies could prevent negative knock-on effects for developing ones—not only by waiting for lower-income countries to introduce their own carbon pricing initiative(s), which will be a challenge given their limited administrative capability, but also by supporting those that need the most help to reduce their emissions. Such support could be provided by dedicating resources and technology to improve the efficiency of industrial processes, financing renewable energy projects, and exempting the poorest countries from the BAM where necessary (Böhringer et al. 2022). The EU and other advanced economies should also dedicate part of the BAM revenue to help developing countries adopt cleaner technologies—to produce greener cement in Vietnam or chemicals in Indonesia, for example—and thus reduce emissions in the long run. By helping to finance the developing world's green transition, developed countries could mitigate the protectionist threat in their own climate agenda.

## Notes

1 Based at the University of Oxford, *Our World in Data* is an online platform of scientific data that focuses on large global problems such as poverty, disease, hunger, climate change, war, existential risks, and inequality (accessed in December 2022).
2 Specifically, trade data were from the Global Trade Analysis Project and production-based estimates of land-use emissions and agricultural products came from Hong et al. (2021), which were based on agricultural emissions from FAOStat provided by the United Nations Food and Agriculture Organization (FAO 2019), and land-use change (LUC) emissions from a spatially explicit bookkeeping model. Emissions were assessed in units of $CO_2$ equivalent with the 100-year warming potentials of $CH_4$ and $N_2O$, and LUC emissions were assigned to the years in which they might have occurred. Readers may also refer to Hong et al. (2022) for the details of the multiregional input-output model used in attributing emissions to a region where the relevant goods are ultimately consumed.

## References

Antweiler, W., Copeland, B.R., Taylor, M.S. 2001. Is free trade good for the environment? *American Economic Review* 91 (4): 877–908.
Babiker, M.H. 2005. Climate change policy, market structure, and carbon leakage. *Journal of International Economics* 65: 421–445.
Böhringer, C., Fischer, C., Rosendahl, K.E., Rutherford, T.F. 2022. Potential impacts and challenges of border carbon adjustments. *Nature Climate Change* 12: 22–29.
Branger, F., Quirion, P. 2014. Would border carbon adjustments prevent carbon leakage and heavy industry competitiveness losses? Insights from a meta-analysis of recent economic studies. *Ecological Economics* 99: 29–39.
Clausing, K., Wolfram, C. 2023. Carbon border adjustments, climate clubs, and subsidy races when climate policies vary. *Journal of Economic Perspectives* 37 (3): 137–162.
Coyle, D. 2022. The double transformations. *Project Syndicate* (December 9).
Durántez, M.G., Obern, C. 2022. The EU's carbon border tax could hurt developing countries. *Project Syndicate* (June 24).
Food and Agriculture Organization of the United Nations (FAO). 2019. *FAOStat*. Available at: www.fao.org/faostat/en/.
Frankel, J. 2022. Let the WTO referee carbon border taxes. *Project Syndicate* (November 29).
Ghertner, D.A., Fripp, M., 2007. Trading away damage: Quantifying environmental leakage through consumption-based, life-cycle analysis. *Ecological Economics* 63 (2-3): 563–577.
Hong, C.P., Zhao, H.Y., Qin, Y., Burney, J.A., Pongratz, J., Hartung, K., Liu, Y., Moore, F.C., Jackson, R.B., Zhang, Q., Davis, S.J. 2022. Land-use emissions embodied in international trade. *Science* 376: 597–603.
Hong, C.P., Burney, J.A., Pongratz, J., Nabel, J.E.M.S., Mueller, N.D., Jackson, R.B., Davis, S.J. 2021. Global and regional drivers of land-use emissions in 1961–2017. *Nature* 589: 554–561.
Hou, F.M., Su, H.Y., Liu, C., Lin, X.X., Zuo, F.Y., Xiao, H. 2022. Carbon emissions embedded in China's paper trade: Estimated outcomes of alternative approaches. *Forest Policy and Economics* 145 (102863).

International Panel on Climate Change (IPCC). 2006. *Guidelines for National Greenhouse Gas Inventories*. Available at: www.ipcc.ch/report/2006-ipcc-guidelines-for-national-greenhouse-gas-inventories/.

International Panel on Climate Change (IPCC). 2020. *Climate Change and Land: Summary for Policymakers*. Available at: www.ipcc.ch/srccl/.

Jing, X.W., Zhao, Q.J. 2019. Research on carbon emission accounting of pulp and paper enterprises based on production process. *China Forest Economics* 6: 9–14 (in Chinese).

Ke, S.F., Qiao, D., Yuan, W.T., He, Y.J. 2020. Broadening the scope of forest transition inquiry: What does China's experience suggest? *Forest Policy and Economics* 118 (102240).

Mehling, M.A., G.E. Medcalf, R.N. Stavins. 2018. Linking climate policies to advance global mitigation. *Science* 359 (6379): 997–998.

Nordhaus, W., 2021. *The Spirit of Green*. New Jersey: Princeton University Press.

Pan, J.H., Phillips, J., Chen, Y. 2008. China's balance of emissions embodied in trade: approaches to measurement and allocating international responsibility. *Oxford Review of Economic Policy* 24 (2): 354–376.

Peters, G.P., Hertwich, E.G. 2008. $CO_2$ embodied in international trade with implications for global climate policy. *Environmental Science & Technology* 42: 1401–1407.

Pietrobelli, C., Staritz, C. 2018. Upgrading, interactive learning, and innovation systems in value chain interventions. *European Journal of Development Research* 30 (03): 557–574.

Ritchie, H. 2019. How do $CO_2$ emissions compare when we adjust for trade? *Our World in Data*. Available at: https://ourworldindata.org/co2/country/china?country=~CHN.

Rodrik, D. 2022. Climate before trade. *Project Syndicate* (December 13).

Stern, N. 2006. *The Economics of Climate Change: The Stern Review*. Cambridge University Press. Available at: http://mudancasclimaticas.cptec.inpe.br/~rmclima/pdfs/destaques/sternreview_report_complete.pdf.

Tong, J.K., Chen, H.Y. 2015. Overview and technical measures of cleaner production in the paper industry. *Resource Conservation and Environmental Protection* 3: 15–22 (in Chinese).

Yan, Y.F., Zhao, Z.X., Wang, R. 2013. A study on China's implied carbon and emission responsibility in foreign trade based on MRIO model. *World Economic Research* 6: 54–58.

Yin, R.S. 2021. Evaluating the socioeconomic and ecological impacts of China's forest policies, programs, and practices: Summary and outlook. *Forest Policy and Economics* 127 (102439).

# 7 Carbon finance and funding for forest sector climate solutions

## Introduction

The objective of this chapter is to address the following two fundamental questions: Given the massive global investment anticipated to limit the catastrophic effects of climate change by drastically reducing greenhouse gases (GHG) emissions, how can the necessary mitigation and adaptation (M&A) actions, including those in the forest sector, be funded, and how have they been actually financed thus far? To appreciate the magnitude of the gross investment required to achieve global "net-zero" emissions of GHG by 2050 in line with the Paris Agreement's climate targets, a recent study by McKinsey & Co. (2022) suggests that an annual global investment of $9.2 *trillion* is needed over the next 30 years. That amount of expenditure, an increase of $3.5 trillion per year from what is spent today (Tyson and Weiss 2022), must be made to transform the energy, transportation, manufacturing, building, and other sectors of the economy as well as to protect and enhance the capacities of carbon capture and storage through land-use and other practices (IPCC 2022).

Therefore, it is worthwhile and timely to inquire into the essentials of carbon finance in general and forest carbon finance in particular. Meanwhile, there has been scant attention devoted to this subject matter. For instance, Parrotta et al. (2022), in their newly released, comprehensive assessment of a decade of REDD+[1] implementation, examined its evolving international finance landscape. Other than lamenting on the slow and limited funding and portraying a not-so-promising outlook, however, the authors were unable to explore forest carbon finance more broadly, let alone the principles and mechanisms of carbon finance. Following an overview of the policies, processes, and rules of the Paris Agreement, including nationally determined contribution (NDC) and carbon accounting, Wang et al. (2021) explored their implications to forest sector climate solutions. But at the end of that article, they noted that "[S]pace limit does not permit us to get into C (carbon) finance, as well as pricing, which we will address in the future."

Building on these and many other studies, this chapter reviews and synthesizes the policy evolution and practical developments of carbon finance

DOI: 10.4324/9781003436652-8

and funding for forest sector climate solutions. I begin with a deliberation of the economic principles of how to reduce GHG emissions and the mechanisms used to achieve that goal through various forms of carbon pricing in the next section. Following that, I overview the recent developments in designing and implementing the carbon pricing mechanisms first, and then examine the roles that public and private sectors can play and have played, beyond explicit pricing, in financing emission reduction and removal (ER&R). My hope is that this broader background will enable readers to form an adequate perspective that can help them approach forest sector climate actions more effectively. I then delve into forest sector actions by examining the current states and relative costs of different initiatives, alternative means of funding, and other relevant issues. Finally, I close this chapter with a summary of the explicit and implicit mechanisms of carbon pricing, and a brief discussion of the potential opportunities for future inquiry and implementation. Whenever necessary, a box or table is included to showcase the advances of a particular mechanism of carbon pricing or to report the relevant financial statistics.

Overall, the existing evidence indicates that only about 25% of global GHG emissions are subject to any price at all, and 75% of emissions that are subject to a price are less than $10 per ton of carbon dioxide equivalent (tCO$_2$e); and REDD+ and market-based forest project/program finance have accounted for a small fraction (~3%) of the total spent on climate mitigation. Moreover, while promising, the more recent developments of carbon pricing remain slow and disappointing, and the financial rewards and penalties continue to be weak in incentivizing polluters to abate their emissions.

Before proceeding, it should be noted that while the cited literature is available from numerous classical and contemporary sources, it is challenging to collate such a vast, dynamic repository of information. Moreover, it is crucial to place the literature in a proper context and shed crucial light on how to satisfactorily confront the challenges we face in climate change M&A. I expect that my efforts of reviewing the policies and practices will contribute to an updated and coherent understanding of the financial complexities of ER&R and an effective implementation of M&A actions defined by the NDCs and other international obligations, including those in the forest sector.

## Principles and mechanisms of carbon pricing

In his book titled *The Spirit of Green*, Professor William Nordhaus (2021) has articulated the economic theory of GHG emission as a global *externality* (or public good) and the mechanisms for climate change mitigation through GHG ER&R. He began by observing that climate change results from the impact of GHG emission that takes place outside the market; thus, GHG emission constitutes a global-scale externality, as the cost it entails (or the benefits of its ER&R) spill outside of the market worldwide and are not directly reflected in prices. For activities associated with this type of broad externality, the costs, benefits, and prices are not properly aligned. Thus, "a well-managed society will ensure that major negative externalities are corrected through government laws that promote negotiations and liability for damage through powers such as regulations and taxes" (p. 19).

He further noted that "the common theme of all externalities is that 'the price is wrong'," meaning that prices do not reflect social costs. In the context of climate change, the *social cost of carbon* (SCC) is the total economic cost incurred for each additional ton of $CO_2$ (or equivalent) emitted to the atmosphere (Nordhaus, 2019). So, he argued that a central principle for dealing with climate change as a grave externality is *federalism*, which "recognizes that legal, ethical, economic, and political obligations and processes operate at different levels, and the solutions will necessarily involve various institutions and decision processes depending on the level" (Nordhaus 2021, p. 22). In fact, many eminent economists agree that the most effective policies are to "internalize" costs and benefits. *Internalization* requires that those who generate externalities pay social costs. "Justice also requires compensation for those who are harmed" (Nordhaus 2021, p. 23). Further, "Societies will deploy different mechanisms to deal effectively with spillovers. These will be market incentives, governmental regulations and fiscal penalties, organizational activities through corporate responsibility, and personal ethics for important interpersonal interactions" (Nordhaus 2021, p. 24).

Indeed, the Paris Agreement has urged national governments and international agencies to gear up the adoption of policies that promote carbon pricing schemes and substantially boost M&A finance (World Bank 2021). As an indispensable part of efficient ER&R, carbon pricing is intended to incentivize changes in investment, production, and consumption patterns and to promote technological innovation and deployment that will drive down future abatement costs. More specifically, scholars have argued that carbon pricing can serve multiple functions (Rogoff 2021, Stern 2006):

- It signals to consumers which products & services are more carbon intensive;
- It offers incentives for investors and innovators to develop new low-carbon technologies;
- It incentivizes producers to adopt low-C practices, processes, and technologies;
- It provides revenues for government and business to invest in climate change M&A.

In practice, carbon pricing can take the form of cap and trade, taxation, corporate internal pricing, and even direct public financing on behalf of the society, although carbon tax and cap-and-trade are the two primary policy mechanisms (Carbon Pricing Leadership Coalition 2017). The keys to any external pricing mechanism are to ensure that the appropriate legal framework exists, public and private sectors are adequately prepared and receptive, and most importantly, the price is appropriately set to drive the desired behavior (Stern 2006).

Before we turn to the different mechanisms of carbon pricing, it is appropriate to note that the Carbon Pricing Leadership Coalition (2017) determined that to achieve the Paris Agreement target—global temperature rise of 1.5°C, GHG emissions should be globally priced at a minimum of $40–80/tCO$_2$e by 2020, ramping up to $50–100/tCO$_2$e by 2030. A recent study, featuring a more pragmatic approach to price carbon in the US by targeting net-zero emissions by 2050, has suggested a similar result: $34–64/tCO$_2$e in 2025 and $77–124/tCO$_2$e in 2030 (Kaufman et al. 2020).

## Putting a price on carbon

Here, I deliberate the specific mechanisms of putting a price on carbon. In addition to carbon tax and emission cap and trade, I also cover voluntary market, corporate internal pricing, and other relevant developments.

### Carbon tax

A carbon tax is a simple and straightforward way to introduce a price for GHG emissions, and each emitter is incentivized to cut emissions to reduce its own tax burden. A carbon tax offers several advantages. First, it puts the burden of GHG emissions on polluters and drives a sharp focus on reducing emissions as well as innovating new technologies (Rogoff 2021). If it is built within the framework of existing tax code, it can be relatively easy to implement and administer. Also, a carbon tax generates revenue that can be used to improve energy efficiency, develop clean or novel technologies like carbon capture and storage (CCS), adapt to climate change effects, and/or fund nature-based solutions (NbSs) (Pettinger 2020).

To facilitate implementation, a carbon tax can be instituted as "revenue neutral," whereby the full revenue stream is returned to businesses or households in the form of rebates or other tax cuts; similarly, public monetary transfers can be used to infuse cash into targeted constituencies that do not contribute significantly to GHG emissions but may experience economic hardship as a result of the carbon tax (Sumner et al. 2009). While the cost of a carbon tax is predictable and can be targeted to specific areas, it does not guarantee achievement of a specific net emission reduction. Further, taxes are notoriously unpopular and politically polarizing. The mere notion of a carbon tax typically evokes stiff resistance from industry and consumers, who generally dislike the idea of taxes and may neither understand nor trust in the notion of revenue neutrality (Pettinger 2020).

Even if public resistance can be overcome, political battles are inevitable as the allocation of tax revenues is decided amongst competing interests. Contentious debate is also likely to arise when considering on whom the carbon tax is to be levied—should it be placed on upstream sources of emissions (producers) or downstream users (consumers)? Taxing upstream users may be easier for government agencies but taxing downstream users may provide a more direct signal to change behaviors (Sumner et al. 2009). Despite its challenges, many economists, politicians, and industry leaders often favor a carbon tax as the most efficient way to drive emission reductions. Several countries, including Sweden, Switzerland, Colombia, and Costa Rica (see Box 7.1) and other jurisdictions, like British Columbia, have successfully reduced emissions through a tax policy, while maintaining robust economic growth (World Bank 2021). Unlike the case of Costa Rica, however, agricultural, forestry, and other land-based sectors in many countries have not directly benefitted from their carbon tax revenues, even though certain activities in these sectors may have been somehow exempted by the policy (Ecosystem Marketplace 2021).

> **Box 7.1 Carbon tax in Costa Rica**
>
> Costa Rica introduced a 3.5% tax on fossil fuels in 1997. Today, proceeds from this carbon tax generate about $27 million a year that is administered by the country's National Forest Fund (FONAFIFO). FONAFIFO supports the conservation of mature forests, reforestation with native species, and agroforestry projects that incorporate a mix of trees with crops or grasslands. In the two decades following the implementation of the tax, Costa Rica's economy has grown an average of more than 4% per year, and the fund has disbursed $500 million to about 18,000 landowners for projects protecting at least one million hectares of mature forest and more than 71,000 hectares of reforestation. Transparency and accountability are keys to the success of government agencies like FONAFIFO, along with its commitment to represent interests of indigenous people, such as the protection of 162,000 hectares of territory occupied by the Cabécar and Bribri tribes (Barbier et al. 2020).

*Cap and trade*

In contrast to a carbon tax, a cap-and-trade system, also known as an emissions trading scheme (ETS), uses market forces to drive emission reductions. While results are mixed, an ETS has the distinct advantage of being more politically and socially acceptable than a carbon tax. Over 80 countries specifically mention carbon markets in their NDCs (Bayer and Aklin 2020).

In an ETS, a governmental body sets an overall limit, or cap, on carbon emissions for a given jurisdiction, and each emitter is granted permits (allowable quantities of emissions) that add up to the cap. Participants are allowed to trade emission credits or debits, provided the total emission cap is achieved for the entire community (Carbon Pricing Leadership Coalition 2017). The key is to regulate a gradual reduction of the cap so that polluters must make commensurate cuts in emissions; as the cap is lowered, scarcity results in an increase in the price of emission allowances. If allowances for an individual firm are unachievable or unaffordable, then that firm can pay another to help achieve its target (Naughten 2011). This is done by purchasing carbon credits from others who operate below their permitted emissions. Participants can also offset emissions by funding projects that sequester and store carbon, such as forest sector NbSs.

The basic premise of an ETS is that it does not matter if every individual polluter reduces emissions equitably, provided that total emissions for the entire jurisdiction fall within the specified cap. The desired way to achieve this objective is for firms to invest in innovation of new technologies or efficiency improvements to reduce their own emissions, and offsets serve as an enabler to help achieve this ultimate objective. In theory, market forces should result in lowest cost of emission reductions, with the idea that benefits will be enjoyed by consumers and society at large (Carbon Pricing Leadership Coalition 2017).

While an ETS induces emissions to align with an overall cap, the price of carbon is difficult to predict as economic growth and other factors cause variability in the trading price. In addition, these schemes often fail to demonstrate measurable ER&R because industries stoke fear of economic calamity and sow doubt to successfully lobby governments to grant generous (even free) permits with limited regulation, oversight, and accountability (Canham 2021). For example, China recently launched the world's largest ETS, representing 12% of global emissions; but due to intense industry lobbying, the initial rollout omits a formal overall emissions cap, basically rendering it ineffective in driving emission reductions (Liu 2021). Also, many ETSs have become exceedingly complicated, often involving speculation in sophisticated financial instruments, like forward derivatives and futures, that benefit an elite class of traders and result in limited impact on actual emissions (Naughten 2011).

Many of these shortcomings can be overcome with an aggressive cap reduction strategy and strong safeguards and enforcement designed to create robust market forces to achieve sustainable ER&R (World Bank 2021). Today, over 30 ETS programs exist around the world, including the European Union, New Zealand, and the state of California (see Box 7.2).

**Box 7.2 Carbon cap-and-trade in California, United States**

In response to the failure of a US national carbon pricing bill in 2003, California spearheaded its own ETS, through Assembly Bill 32 (AB-32), in 2006. AB-32 was designed to reduce GHG emissions to baseline (1990) levels by 2020 (realized in 2016) and further reduce emissions by 40% and 80% below baseline by 2030 and 2050. California also intends to generate 100% of its electricity carbon-free and achieve economy-wide carbon neutrality by 2045. Since 2015, 85% of the state's GHG emissions have been governed by AB-32. The overall cap decreased by 3% per year between 2015 and 2020 and is designed to decrease another 5% per year by 2030. As a result,

AB-32 contributed to a 5.3% GHG emission reduction in 2013–2017 and generated $5 billion of revenue, 35% of which was invested in low-income and environmentally disadvantaged communities (Center for Climate and Energy Solutions 2022).

To incentivize GHG emission reductions and carbon sequestration, California Air Resources Board (CARB) issues offset credits to qualified projects pursuant to six Board-approved Compliance Offset Protocols (on livestock, ozone depletion substance, US forests, urban forests, mine methane capture, and rice cultivation projects). These offsets are tradable credits that represent verified emissions reductions or removal enhancements from sources not subject to a compliance obligation in the cap-and-trade program. Compliance entities may use offset credits to meet up to 8% of their obligation for emissions through 2020; 4% of their obligation for emissions from 2021–2025; and 6% for emissions from 2026–2030. Forestry credits in the offset market surged to more than 83 million tons of $CO_2e$ during 2017–2019, valued at almost $1.2 billion (CARB 2021). However, a recent study has reported that carbon accumulated in offset projects has not been additional to what might have otherwise occurred (Coffield et al. 2022).

*Voluntary markets*

With increasing pressure on companies to adopt "net-zero" emission goals as part of their corporate social responsibility (CSR), voluntary markets are becoming popular and evolving rapidly. Like compliance markets, voluntary markets provide an alternative path for organizations to offset GHG emissions through the purchase and sale of carbon credits. But as the name implies, these are neither mandated nor regulated (World Bank 2021). Nevertheless, voluntary markets are gaining credibility as large, reputable organizations like Microsoft and Amazon commit to net-zero targets, and American Carbon Registry, Gold Standard, and Verra Carbon Standard, among others, develop protocols for certifying ER&R projects (DeFries et al. 2022, Ecosystem Marketplace 2021).

It is important to emphasize that effective corporate environmental, social, and governance (ESG) strategies strive for true emission reductions at the source. Unfortunately, in pursuit of "quick wins," some companies may be tempted to purchase carbon offsets, including those predicated on REDD+ projects, as a substitute for reducing their own emissions (Parrotta et al. 2022). While initially attractive, this approach is insufficient to reduce GHG emissions in the spirit of the Paris Agreement; offsets should only be used to complement investments that abate emissions (World Bank 2021). Further, socially responsible companies look to optimize their full value chains by selecting upstream partners with low GHG emissions and/or supporting supply partners to achieve emission reductions. When used properly, the purchase of carbon credits offers a great opportunity for companies to realize net-zero emissions across their entire value chains, while providing a funding source for offsetting projects, such as NbSs.

Another question is how to ensure the trustworthiness of the large intermediaries, who are involved in the project development, verification, and certification, and thus the credibility of the transactions over voluntary markets. Having brought up the findings of a recent investigative report by *The Guardian* in Chapter 2, I remarked on the acute need for adequate supervision and regulation by certifying the certifiers and verifying the verifiers as part of the discussion over alternative approaches to governing climate solutions. Along the way, I also suggested the relevance to integrate voluntary markets and other carbon projects and programs of the private sector into a national or subnational jurisdiction. Hopefully, these ideas will attract the public attention, and the offset credits generated by voluntary markets will be robust and reliable.

## Other financing sources

There has been growing recognition that capital in global markets must be shifted toward low-carbon activities. Effective public policy can induce research and development of new technologies to reduce emissions and should foster investments in such areas as renewable energy, increased efficiency standards, public transportation, urban planning, and land management. One could argue that current global energy policy, in the form of subsidies to the fossil fuel industry, has the opposite effect by promoting GHG emissions. Data for 2020 indicate that of the $634 billion in global energy sector subsidies, about 70%, or nearly $450 billion went to fossil fuels (Taylor 2020). Clearly, this disparity must be addressed where it exists, partly by *phasing out the fossil fuel subsidies.*

On the other hand, certain jurisdictions are reluctant or unable to impose costs of ER&R directly on firms and consumers, which can lead to either inaction or to imposition of costs in other forms. A prominent example is that the US federal government has recently enacted a burst of *public spending on clean energy and innovation* in 2021 and 2022, including funding for clean energy infrastructure and investments as well as a long list of clean energy tax credits (Clausing and Wolfram 2023). It has been reported that the 2022 Inflation Reduction Act alone would spend more than $350 billion over ten years on clean energy tax credits and subsidies (Congressional Budget Office 2022).

*Climate or "green" bonds* involve the use of fixed-income instruments designed to finance debt for climate solutions (World Bank 2014). The global green bond market has grown from about $13 billion in 2013 (World Bank 2014) to over $646 billion in 2021 and is expected to surpass $1 trillion in 2022 (Jones 2022). However, bonds have limitations. Investors are not given a guarantee that their money will be spent directly on a specific project of their desire. Budget constraints or scarcity of capital may force debtors to invest in alternative priorities. Despite these and other challenges, green bond investment continues to grow because it allows organizations to signal to investors their commitment to sustainability. Moreover, these bonds also offer investors an opportunity to diversify portfolios with an ESG agenda (Giugale 2018). While not yet a significant source of capital for forest projects, global bond proceeds support projects such as renewable energy, transportation, building energy efficiency, land use, and the like (Jones 2022).

The Carbon Pricing Leadership Coalition (2017) observed that "Carbon pricing can also be implemented by embedding notional prices in, among other things, financial instruments and incentives that foster low carbon programs and projects." One way this is done is for companies to institute their own *internal carbon price* on carbon. This can be accomplished with an *internal carbon fee* (monetary value on carbon emissions used to generate a dedicated investment stream for ER&R projects); a *shadow price* (internal hurdle rate designed to support investments and create an internal inducement to decarbonize); or an *implicit price* (spotlights the amount spent to comply with efficiency standards or reduce emissions to alleviate carbon footprint). Regardless, the aim is to place a monetary value on GHG emissions and factor the associated costs into existing operations and future investments. Many companies also advocate this approach to help build more resilient supply chains and support CSR objectives by responding to shareholder concerns. According to the Center for Climate and Energy Solutions (2022), over 1,200 companies have either instituted or plan to implement internal carbon pricing.

In addition to domestic public and private finance in developed countries, political leaders and academic experts have stressed the urgency of *international finance* to "assist developing countries in mitigating the effects of climate change without further exacerbating their already-unsustainable debt levels" (Rogoff 2023). Some have even called for the establishment of a "global green bank" or "global carbon bank," which Rogoff (2023) argues is a proposal "that rich countries must consider if they are serious about tackling climate change… and promoting prosperity … in the developing world." More broadly, just like public-private partnership can help drive innovation and investment for a sustainable future (Tyson and Weiss 2022), international cooperation and linkage will reduce distortions in trade and capital flows and aid the efficient ER&R by promoting consistency of actions across countries (Nordhaus 2021).

In short, in the spirit of pursuing jurisdictional approaches to governing climate solutions, there is a significant role for public policy and government action to foster investment, and complementary roles for the private sector, multilateral development banks, international financial institutions, and concessional finance of various forms (Songwe et al. 2022).

**Forest sector solutions**

World forests are huge carbon sinks, absorbing ~15.6 billion tons of $CO_2$ per year; because of deforestation and other disturbances, however, they emit about half of that quantity back into the atmosphere (Harris et al. 2021).

122  *Carbon finance and funding*

Thus, deforestation and forest degradation remain a great source of $CO_2$ emissions. Altogether, 23% of anthropogenic GHG emissions currently come from the destruction or poor management of the world's forests, farms, and pasture lands (Ecosystem Marketplace 2021). On the other hand, forest sector actions could offer over two-thirds of the cost-effective NbSs necessary to hold warming to below 2.0°C and about half of the low-cost mitigation opportunities (Griscom et al. 2017).

The argument for using forest sector NbSs to combat climate change is compelling. Stern (2006) observed that reducing deforestation is the "single largest opportunity for cost-effective and immediate reductions of carbon emissions." Indeed, it is estimated that about 14% of all emission reductions required to achieve net-zero GHG emissions by 2050 could be realized at a relatively low cost (~$10/tCO$_2$e) by eliminating deforestation (McKinsey & Co. 2022). Ironically, forest ecosystem conservation only receives a small fraction (~3%) of the total spent on climate mitigation (Barbier et al. 2020).

## *Supporting REDD+*

The Paris Agreement "recognizes the importance of adequate and predictable financial resources … for reducing emissions from deforestation and forest degradation," while affirming the significance of non-carbon co-benefits, such as community livelihoods and biodiversity conservation (UNFCCC 2020). Many developing countries rely on REDD+ activities in their NDCs to achieve emission reduction targets (Duchelle et al. 2017). While effective policy should be shaped and enacted by the nations where deforestation occurs, many poor countries lack the economic resources to fund the effort (Parrotta et al. 2022). As a result, industrialized countries pledged at COP15 $100 billion a year by 2020 to help developing countries cut carbon emissions and deal with the effects of climate change. To administer this fund, the Green Climate Fund (GCF) was established as the financial arm of the UNFCCC with a mandate to invest equally between M&A. Unfortunately, that pledge has not been delivered (Roberts et al. 2021). Despite a flurry of promises leading up to COP26 in Glasgow, only about $80 billion a year was committed in 2021, most in the form of public grants or loans. Also, only about 25% of the funding has been designated for adaptation projects, falling short of the GCF mandate (Timperley 2021).

In an encouraging sign, over 100 world leaders at COP26 committed to end deforestation by 2030, including a total pledge of $19.2 billion in public and private funds (Parrotta et al. 2022). Notably, the 2021 pledge includes key players like Brazil, Russia, and China, who did not sign the 2014 declaration. In addition, the US also committed $9 billion to conserve and restore the world's forests. These commitments are critical, and the relatively low cost and high impact of REDD projects are borne out by the data. During its initial five-year pilot (2014–2018), for instance, eight REDD projects (mostly in the Amazon) totaling more than 100 million tCO$_2$e in emission reductions were funded for less than $500 million, or at a cost about $5/tCO$_2$e (Green Climate Fund 2022).

*Market-based funding*

Forest-based carbon projects may also be funded through the sale of offsets in compliance and voluntary markets. In either case, the trading of carbon credits creates two desired effects: First, market forces are leveraged to drive efficient emission reductions; second, essential investments in NbSs are made. In market transactions, the process normally begins with a forest owner (land manager) who authors a project design document (PDD) describing how the project will lead to ER&R, which may take the form of afforestation and reforestation (A&R) or improved forest management (IFM), as well as REDD. The PDD includes a comparison of a "business as usual" (BAU) base case estimate of emissions if the project was not implemented, compared with predicted emissions resulting from the project. It is then submitted for review under a rigorous protocol designed to verify ER&R before the project can be certified and offset credits issued. Various registries are engaged in certifying projects in voluntary markets, whereas the CARB and the Regional Greenhouse Gas Initiative are examples of regulatory agencies that certify projects in compliance markets (Elyse et al. 2018).

Critics point out that the process to develop offsets is slow and may overstate a project's impact, especially in voluntary markets, due to one or more of the following factors:

1   the "additionality" or true net impact on ER&R may be overestimated because of inaccurate BAU emissions that are derived from unrealistic scenarios (Canham 2021, Coffield et al. 2022);
2   ER&R from protecting, rehabilitating, or reforesting one area can leak to another area and result in forest degradation and deforestation or A&R that may not have occurred otherwise (DeFries et al. 2022); and
3   inadequate or inaccurate measurement, reporting and verification (MRV) of carbon emissions from a project may exaggerate its benefit.

(Naughten 2011)

While there are active and promising trends in these markets, the overall market size and trading price remains relatively small.

*Jurisdictional finance*

To address the inefficiency and limited capability associated with the project-oriented approach and to accelerate the pace of forest sector M&A, governments, businesses, and other players ("donors") have been increasingly engaged in carbon financing through *jurisdictional* approaches. In this case, the donor—who wishes to reduce emissions via REDD or to enhance removals via A&R or IFM—finances forest sector NbSs across an entire jurisdiction instead of for a single project (Ecosystem Marketplace 2021). The resulting payments are made to the jurisdictional authorities where the NbSs are pursued.

As a core component of the climate change M&A governance paradigm under the Paris Agreement (Wang et al. 2021), jurisdictional funding represents an important turning point in carbon financing because until recently, most forest results-based finance has involved market-based funding, while the designated public-sector funding has been funneled to REDD readiness efforts (Ecosystem Marketplace 2021). More importantly, countries have undertaken ecological restoration programs (ERPs) as part of their commitments to sustainable development. While many of these ERPs may not have been originally conceived for carbon sequestration or emission offsetting, they are now called upon to assume this additional responsibility (IPCC 2022). With a general compatibility between carbon sequestration and storage and other ecosystem services, and the newly gained currency following the UN declaration of this decade as the Decade of Ecosystem Restoration (Cook-Patton et al. 2021), ERPs have become a "game changer" for funding ER&R. Let me summarize the Chinese experience in Box 7.3.

**Box 7.3 China's ecosystem restoration and conservation**

The Chinese government invested about 700 billion yuan ($110 billion) by 2017 in several large ERPs, such as the Sloping Land Conversion Program, the Natural Forest Protection Program, the Three-Norths Shelterbelt Program, the Beijing-Tianjin-Hebei Desertification Combating Program, and the Wildlife Protection and Nature Reserve Program (NFGA 2018). These programs have greatly improved the ecological conditions and the local people's livelihoods (Yin 2009). Also, the country has planned to spend about three trillion yuan ($470 billion) on nine large regional ecosystem restoration and conservation initiatives during 2021–2035, which are expected to further improve the ecological condition and the local people's livelihoods (NFGA 2022). When these programs/initiatives were originally conceived and planned, however, carbon sequestration and storage by forest and other terrestrial ecosystems was not explicitly considered. With the country's updated NDC targets to peak GHG emissions this decade and reach carbon neutrality before 2060, nonetheless, they have attracted a great deal of attention for removing GHGs emitted into the atmosphere and thus partially offsetting its gross emissions (Hou and Yin 2022).

Jurisdictional funding has also opened new opportunities for "nesting" and "hybridizing" market-based, public, and other private alternatives; thus, it offers distinct advantages that are expected to accelerate funding of forest sector NbSs (Seymour 2020). For example, unlike project developers, governments have the authority to enforce policy and control land use change broadly, while the private sector can serve as a source of immediate results-based payments. Also, incentives for aggregating projects across a jurisdiction can mitigate the risks of non-additionality, non-permanence, and leakage, and the threats to indigenous rights (van der Gaast et al. 2017). Finally, as elaborated in Chapter 2, this approach can reduce the transaction costs and risks associated with individual projects.

Table 7.1 shows that REDD+ and market-based forest project/program finance is ramping up in recent years. As indicated by the 2017–2019 data, about 39% of money spent on forest projects were REDD+ funded through the GCF, while 52% of global financing came through compliance markets and only about 9% came from voluntary markets—too little to make a big difference. Not only are these categories expected to surge in the next decade, but it is also believed that jurisdictional finance will emerge rapidly as public pressure to achieve carbon neutrality and ecosystem restoration mounts (Ecosystem Marketplace 2021).

## Summary and outlook

This chapter was motivated to better understand the carbon pricing and financing mechanisms that are not just currently available but also required for achieving ER&R in line with the Paris Agreement climate targets. I began with an overview of the economic principles of reducing GHG emissions and the mechanisms of carbon pricing. Then, after introducing those explicit and implicit mechanisms, I elucidated the roles that the private and public sectors can play and have played in financing ER&R. I also scrutinized the current states of forest sector M&A actions, alternative ways they are funded, and their efficiency and effectiveness.

To sum up, although carbon pricing alone may not be adequate to induce change at a pace or on the scale required to meet the Paris Agreement temperature targets, it is vital to create a strong incentive to drive down GHG emissions (Nordhaus 2021, Stern 2006), which can take a variety of forms and must be complemented by well-designed policy (World Bank 2021). The concrete mechanisms—carbon tax, compliance and voluntary emission

Table 7.1 Major categories of forest carbon finance

| Category | 2017–2019 | Since 2009 |
| --- | --- | --- |
| GCF funded REDD | $1,717 million ($400 million from results-based payments) | $1,935 million ($619 million from results-based payments) |
| Compliance market offsets | $2,334 million | $3,908 million |
| Voluntary market offsets | $397 million | $1,394 million |

Note: Data came from Ecosystem Marketplace (2021).

trading, internal pricing, or financing via issuing bonds or mobilizing public budgetary resources—are summarized in Table 7.2. Regardless, however, it is prudent to note that we are far from the envisioned extent of carbon pricing and level of carbon price. Only about 25% of global GHG emissions are subject to any price at all, and 75% of emissions that are subject to a price are less than $10 per $tCO_2e$ (World Bank 2021). Furthermore, forest sector M&A actions are rarely linked up with or integrated into these pricing mechanisms. While promising, the more recent developments of carbon pricing remain slow and disappointing and the financial rewards and penalties continue to be weak in incentivizing polluters to abate their emissions (World Bank 2021).

The more seamlessly and comprehensively carbon pricing is introduced on a global scale and the more aggressively pricing is ratcheted up, the more effective it will be in limiting the devastating impacts of climate change (Carbon Pricing Leadership Coalition 2017). But care must be taken to ensure that the introduction of carbon pricing is done in a way that not only limits emissions and potential climate disruption, but also fosters economic growth and social-ecological sustainability (DeFries et al. 2022). Fortunately, there are many pricing mechanisms that have been successfully deployed to achieve these objectives, and novel approaches continue to emerge as different mechanisms are strengthened, nested, hybridized, and/or integrated. A combination of them is likely to be more efficient, effective, and even equitable (World Bank 2021). In any case, market mechanisms can and must be harnessed; but "they will succeed only if the governance principles and operating methods are designed to promote equity and sustainable prosperity" by aligning financial markets with climate commitments (Zadek 2023).

Finally, I highlighted the opportunities of forest sector climate solutions, including REDD+, as well as their significance and comparative advantage. Clearly, there are many other land-use practices that can contribute to GHG ER&R, including grassland, wetland, peatland conservation or restoration, climate-smart agriculture and agroforestry, soil conservation, and coastal restoration. These efforts should also be part of the ultimate NbSs (IPCC 2022, Sato et al. 2019). But for any land-use practices to be effective and sustainable, we must confront the risks of potential non-additionality, non-permanence, leakage, and high transaction costs resulting from individual projects (DeFries et al. 2022, Parrotta et al. 2022). Whenever feasible, some kind of pooling or aggregation via intermediary agencies or jurisdictional organizations should be deployed; accordingly, regional, program-based approaches beyond the project orientation is called for (Wang et al. 2021). As such, the effects of mitigating project-level risks and costs of these alternative funding and participating pathways, as well as their effectiveness, deserve closer scrutiny.

Table 7.2 Carbon pricing alternatives

| Sector | Mechanisms | Benefits | Concerns |
|---|---|---|---|
| Public | Carbon tax | • Simple & easy to administer<br>• Burden is on polluters<br>• Generates revenue that can target | • Politically unpopular<br>• Does not ensure ER&R targets<br>• Allocation of revenue is divisive |
|  | Compliance ETS | • Leverages market forces<br>• Caps emissions<br>• Politically & socially accepted<br>• Aligned with NDCs | • Unpredictable C price<br>• Mixed success due to special interest influence<br>• Complexity may exclude all but elite traders |
|  | Regulatory action (energy subsidies, emission standards, urban planning, etc.) | • Policy framework already exists<br>• High potential impact | • Politically charged<br>• Difficult to adapt with changing conditions |
| Private &Public | Jurisdictional finance | • Governments enforcement<br>• Mitigates risk of non-additionality or leakage<br>• Consistent with UNFCCC/PA | • Relatively new and untested<br>• Undeveloped certification guidance |
| Private | Internal C pricing (shadow price, implicit price, etc.) | • Internal investment control<br>• Sustainable business growth<br>• Support corporate social responsibility | • Limited consistency and accountability<br>• Can reduce short-term profitability |
|  | Voluntary markets | • Consistent with capitalist economies<br>• Enables companies with low emissions to achieve C neutrality | • Inconsistent certification quality of offsets<br>• May encourage avoidance of emission reductions |
|  | Bond funding | • Company strongly reinforces commitment to sustainability<br>• Offers diversification for ESG investors | • Funding may not go to desired projects<br>• Impact often difficult to measure<br>• Budgetary trade-offs can cause conflict of priorities |
|  | Divestment strategy | • Investments can be diverted from polluting companies or industries | • Indirect leverage that does not guarantee results |

## Note

1 See Introduction for a definition of REDD+.

## References

Barbier, E., Lozano, R., Manuel-Rodriguez, C., Troeng, S. 2020. Adopt a carbon tax to protect tropical forests. *Nature* (February 12). Available at: www.nature.com/articles/d41586-020-00324-w?proof=t.

Bayer, P., Aklin, M. 2020. The European Union Emissions Trading System reduced CO2 emissions despite low prices. *PNAS* 117 (16): 8804–8812.

California Air Resources Board (CARB). 2021. *California's Compliance Offset Program*. Available at: ww2.arb.ca.gov/sites/default/files/2021-10/nc-forest_offset_faq_20211027.pdf.

Canham, C. 2021. Rethinking forest carbon offsets. *Cary Institute of Ecosystem Studies*. Available at: www.caryinstitute.org/news-insights/feature/rethinking-forest-carbon-offsets.

Carbon Pricing Leadership Coalition. 2017. *Report of the High-Level Commission on Carbon Prices*. Washington DC: World Bank Group.

Center for Climate and Energy Solutions. Available at: www.c2es.org/.

Clausing, K., Wolfram, C. 2023. Carbon border adjustments, climate clubs, and subsidy races when climate policies vary. *Journal of Economic Perspectives* 37 (3): 137–162.

Coffield, S.R., Vo, C.D., Wang, J.A., Badgley, G. et al. 2022. Using remote sensing to quantify the additional climate benefits of California forest carbon offset projects. *Global Change Biology* 28: 6789–6806. Also available at: https://doi.org/10.1111/gcb.16380.

Congressional Budget Office. 2022. *Congressional Budget Office Cost Estimate*. Available at: www.cbo.gov/system/files/2022-08/hr5376_IR_Act_8-3-22.pdf.

Cook-Patton, S.C., Shoch, D., Ellis, P.W. 2021. Dynamic global monitoring needed to use restoration of forest cover as a climate solution. *Nature Climate Change* 11: 366–368.

DeFries, R., Ahuja, R., Friedman, J., Gordon, D.R., Hamburg, S.P., Kerr, S., Mwangi, J., Nouwen, C., Pandit, N. 2022. Land management can contribute to net zero. *Science* 376 (6598): 1163–1165.

Duchelle, A.E., C. de Sassi, P. Jagger, M. Cromberg, A.M. Larson, W.D. Sunderlin, S.S. Atmadja, I.A.P. Resosudarmo, and C.D. Pratama. 2017. Balancing carrots and sticks in REDD+: implications for social safeguards. *Ecology and Society* 22 (3): 2. https://doi.org/10.5751/ES-09334-220302.

Ecosystem Marketplace. 2021. *State of Forest Carbon Finance 2021*. Washington, DC: Forest Trends Association.

Elyse, C., Kerr, A., Morton, S., Seal, A., Voehler, K., Yan, L., Zayamandakh, U., Reibstein, R. 2018. *Forest Carbon Credits: A Guidebook to Selling Your Credits on the Carbon Market*. Department of Earth and Environment, Boston University.

Food and Agriculture Organization of the United Nations (FAO). 2016. *Forestry for a Low-Carbon Future: Integrating Forests and Wood Products in Climate Change Strategies* (FAO Forestry Paper 177). Rome, Italy.

Giugale, M. 2018. The pros and cons of green bonds. *The World Bank OMFIF Bulletin* (September).

Graham, P., 2015. Forests at UNFCCC-COP21: Brief. *WWF Forest and Climate Programme*. Available at: https://wwfint.awsassets.panda.org/downloads/forest_basics_brief_cop21.pdf.

Green Climate Fund. 2022. Available at: www.greenclimate.fund/redd#projects.
Griscom, B.W., Adams J., Ellis P.W., Houghton R.A., Lomax G., Miteva D.A., Schlesinger W.H., Shoch, D. et al. 2017. Natural climate solutions. *PNAS* 114 (44): 11645–11650.
Harris, N. and Gibbs, D. 2021. *Forests Absorb Twice as Much Carbon as They Emit Each Year.* World Resources Institute. Available at: www.wri.org/insights/forests-absorb-twice-much-carbon-they-emit-each-year.
Hou, J.Y., Yin, R.S. 2022. How significant a role can China's forest sector play in decarbonizing its economy? *Climate Policy* 23 (2): 226–237. doi:10.1080/14693062.2022.2098229.
IPCC. 2020. *Climate Change and Land: Summary for Policymakers.* Available at: www.ipcc.ch/srccl/.
IPCC. 2022. *Climate Change 2022: Impacts, Adaptation and Vulnerability.* Working Group II Contribution to the Sixth Assessment Report.
Jones, L. 2022. $500 bn Green Issuance 2021: social and sustainable acceleration: Annual green $1tn in sight: Market expansion forecasts for 2022 and 2025. *Climate Bonds Initiative* (January 31, blog post).
Kaufman, N., Barron, A.R., Krawczyk, W. et al. 2020. A near-term to net zero alternative to the social cost of carbon for setting carbon prices. *Nature Climate Change* 10: 1010–1014. https://doi.org/10.1038/s41558-020-0880-3.
Liu, H.Q. 2021. In-depth Q&A: Will China's emissions trading scheme help tackle climate change? *Carbon Brief.* Available at: www.carbonbrief.org/in-depth-qa-will-chinas-emissions-trading-scheme-help-tackle-climate-change.
McKinsey & Co. 2022. *The Net-Zero Transition: What It Would Cost, What It Would Bring.* McKinsey Global Institute (January).
Naughten, A. 2011. *Designed to Fail? The Concepts, Practices and Controversies Behind Carbon Trading.* Available at: www.fern.org/fileadmin/uploads/fern/Documents/FERN_designedtofail_internet_0.pdf.
National Development and Reform Commission (NDRC). 2022. *A National Strategic Plan of Ecosystem Restoration and Conservation over 2021–2035.* Beijing, China.
National Forestry and Grassland Administration of China (NFGA). 2018. *China Forestry Yearbook (2017).* Beijing, China.
Nordhaus, W. 2019. Climate Change: The ultimate challenge for economics. *American Economic Review* 109: 1991–2004. Available at: https://doi.org/10.1257/aer.109.6.1991.
Nordhaus, W. 2021. *The Spirit of Green.* New Jersey: Princeton University Press.
OECD (Organisation for Economic Co-operation and Development). 2021. *Carbon Pricing in Times of COVID-19: What Has Changed in G20 Economies?* Paris: OECD. Available at: www.oecd.org/tax/tax-policy/carbon-pricing-in-times-of-covid-19-what-has-changed-in-g20-economies.htm?_ga=2.133364810.515584805.1649599297-1674718103.1649599297.
Ostrom, E. 2010. Beyond markets and states: Polycentric governance of complex economic systems. *American Economic Review* 100: 641–672.
Parrotta, J., Mansourian, S., Wildburger, C., Grima, N. (eds). 2022. *Forests, Climate, Biodiversity and People: Assessing a Decade of REDD+.* IUFRO World Series Volume 40. Vienna.
Pettinger, T. 2020. Carbon Tax – Pros and Cons. *Economics.help.* Available at: www.economicshelp.org/blog/2207/economics/carbon-tax-pros-and-cons/.
Roberts, J.T., Weikmans, R., Robinson, S., Ciplet, D., Khan, M., Falzon, D. 2021. Rebooting a failed promise of climate finance. *Nature Climate Change* 11: 180–182.

Rogoff, K. 2021. Will this COP be different? *Project Syndicate* (November 1).

Rogoff, K. 2023. Rethinking climate finance for the developing world. *Project Syndicate* (June 29).

Sato, I., Langer, P., Stolle, F. 2019. *Enhancing NDCs: Opportunities in the Forest and Land-Use Sector*. World Resources Institute (November).

Seymour, F. 2020. *INSIDER: 4 Reasons Why a Jurisdictional Approach for REDD+ Crediting I Superior to a Project-Based Approach*. World Resources Institute (May).

Songwe, V., Stern, N., Bhattacharya, A. 2022. *Finance for Climate Action: Scaling Up Investment for Climate and Development*. London: Grantham Research Institute on Climate Change and the Environment, London School of Economics and Political Science.

Stern, N. 2006. *The Economics of Climate Change: The Stern Review*. Cambridge University Press. Available at: http://mudancasclimaticas.cptec.inpe.br/~rmclima/pdfs/destaques/sternreview_report_complete.pdf.

Sumner, J., Bird, L., Smith, H. 2009. *Carbon Taxes: A Review of Experience and Policy Design Considerations*. National Renewable Energy Laboratory, Technical Report NREL/TP-6A2-47312.

Taylor, M. 2020. *Energy Subsidies: Evolution in the Global Energy Transformation to 2050*. International Renewable Energy Agency.

Timperley, J. 2021. The broken $100-billion promise of climate finance—and how to fix it. *Nature* 598, 400–402. https://doi.org/10.1038/d41586-021-02846-3.

Tyson, L., Weiss, D. 2022. Public-private decarbonization. *Project Syndicate* (April 22).

United Nations Framework Convention on Climate Change (UNFCCC) Secretariat. 2020. *Reference Manual for the Enhanced Transparency Framework under the Paris Agreement*. Bonn, Germany.

Van der Gaast, W., Sikkema, R., Vohrer, M. 2018. The contribution of forest carbon credit projects to addressing the climate change challenge. *Climate Policy* 18: 42–48.

Wang, Y., Li, L., Yin, R.S. 2021. A primer on forest carbon policy and economics under the Paris Agreement: Part I. *Forest Policy and Economics* 132 (102595).

World Bank. 2014. *Financial Institutions Taking Action on Climate Change*. World Bank Group and United Nations Environmental Programme.

World Bank. 2021. *State and Trends of Carbon Pricing*. Washington, DC.

Yin, R.S. 2009. *An Integrated Assessment of China's Ecological Restoration Programs*. Dordrecht, The Netherlands: Springer.

Zadek, S. 2023. A new playbook for preserving nature. *Project Syndicate* (July 25).

# 8 Risk, uncertainty, and social discounting in coping with climate change

## Introduction

In *The Economics of Climate Change* (also called the *Stern Review*), Professor Stern (2006) succinctly expressed that

> Climate change presents a unique challenge for economics: it is the greatest and widest-ranging market failure ever seen. The economic analysis must therefore be global, deal with long time horizons, have the economics of risk and uncertainty at centre stage, and examine the possibility of major, non-marginal change.

It is indeed important for scholars of forest economics, policy, and finance to incorporate risk and uncertainty properly into their inquiry of how the forest sector can mitigate and adapt to climate change over the long term, in addition to coping with other challenges, some of which I address elsewhere in this book.

Of course, this does not imply that these scholars have failed to incorporate any risk and uncertainty into their economic, policy, and/or financial analyses in the past. Rather, it means that what forest economics, policy, and finance scholars have done so far may not be appropriate or adequate. For instance, studies have examined market volatility and tenure insecurity when determining the rotation age and land value of a timber stand (e.g., Mendelsohn 1994, Yin and Newman 1997, Ning and Sun 2019). These and other similar efforts are well-justified and meaningful if we consider timber as the dominant output in a world without climate change. Once our attention is focused on the role of carbon sequestration and storage by the forest sector, however, we enter a drastically different world whereby we face the overarching question of how to mitigate climate change over a very long horizon. As a result, the frameworks and tools we use to deal with risk and uncertainty previously could have become obsolete or even irrelevant (Stern 2006, Nordhaus 2019). Therefore, it is crucial to rethink what risk and uncertainty we need to include in our future work and how to accomplish this goal to our satisfaction.

More recently, analysts have also paid attention to the socioeconomic impacts of such forest disturbances as fires, pest outbreaks, diebacks, and

DOI: 10.4324/9781003436652-9

hurricanes, which have been intensifying because of climate change (IPCC 2022a). For example, Henderson et al. (2022) estimated the market and welfare impacts caused by Hurricane Michael in the region of Florida's Panhandle and adjacent areas of Alabama and Georgia. Based on forest inventory and remotely sensed data, they assessed the damages to forest area and growing stock and then showed that timber prices declined initially, rose once the demand returned to normal, and finally converged to the baseline trend. Compared with the no-hurricane scenario, the welfare gain for pine sawtimber producers increases by 1.2 to 1.5 times and that of consumers dropped to 0.6 to 0.8 times; likewise, the hardwood sawtimber producers gained 1.8 times in welfare, while consumers lost a half. Similarly, employing a computable general equilibrium model, Patriquin et al. (2007) evaluated the regional economic impacts of the mountain pine beetle outbreak in five areas of British Columbia, Canada. Their results revealed a relative boom to regional economies in the short run because of the increased volume of available timber but a negative impact in the long run because the timber supply fell below the baseline level.

As noted by the IPCC (2020), however, climate change can exacerbate land degradation processes, including through increases in rainfall intensity, flooding, drought frequency and severity, heat stress, dry spells, wind, sea-level rise and wave action, and permafrost thaw with outcomes being modulated by land management. Thus, while it is beneficial to conduct *ex post* assessments of the socioeconomic impacts of various forms of forest disturbances, it is essential to capture the ecological impacts of these disturbances in *ex post* assessments as well. Moreover, it is equally important to carry out integrated economic and ecological assessment of forest disturbances *ex ante* by projecting what damages might happen to the ecological and human systems under given scenarios of climate change (IPCC 2022a). In this context, risk and uncertainty are more about the potential damages caused by climate change (IPCC 2022a).

Furthermore, the combination of these factors—coping with likely severe damages of climate change over a long horizon—forces us to confront the challenge of social discounting in assessing the costs and benefits of climate change mitigation (and adaptation). It has been recognized that when we consider investing in climate change mitigation, a normal market-based discounting scheme of commercial investment is not appropriate; indeed, even a modestly high but constant discount rate used in public finance may no longer be suitable, either (Weitzman 1998, Gollier 2002, Stern 2006). We must think through how we discount the future benefits and costs more carefully: Should we adopt a high or low rate, and should it be constant or declining over time? Also, given our preference for climate solutions sooner rather than later, should we put a time value on the physical ER&R that the forest sector delivers? Even within the realm of nature-based solutions (NbSs) to the climate crisis (Roe et al. 2019, Griscom et al. 2017), questions exist regarding how they should be included in the portfolio of mitigating actions, to what extent they can be financed through different carbon pricing mechanisms (Carbon Pricing Leadership Coalition 2017), and so on.

Obviously, these are some tough questions, to which we may not have satisfactory answers yet. Nonetheless, we must do what we can to situate our research in the proper context and examine them with sound frameworks and suitable tools. Without a complete and coherent understanding of these questions, the significance and impact of our work would be diminished. The objective of this chapter is to explore how scholars of forest economics, policy, and finance can effectively tackle issues related to the risk and uncertainty induced by climate change and those entailed in discounting the future costs and benefits of forest sector climate mitigation. The chapter will unfold in the following four interrelated sections:

1 identifying and dealing with risk and uncertainty;
2 understanding social discounting and its linkage with the social cost of carbon;
3 envisioning the finance of climate mitigation actions, including forest sector NbSs; and
4 embracing the call for a time value of physical carbon.

In writing this chapter, I have benefitted immensely from referencing a large body of literature, with information drawn particularly from Weitzman (1998, 2001), Stern (2006), Gollier and Hammitt (2014), Nordhaus (2019), and Rennert et al. (2021), in addition to the IPCC Assessment Reports.

## Identify and address more relevant risks

Professor Stern (2006) noted:

> No one can predict the consequences of climate change with complete certainty; but we now know enough to understand the risks.... Averaging across possibilities conceals risks. The risks of outcomes much worse than expected are very real and they could be catastrophic. Policy on climate change is in large measure about reducing these risks. They cannot be fully eliminated, but they can be substantially reduced.
>
> (p. 9)

Across the IPCC working groups of the Sixth Assessment Report,

> risk provides a framework for understanding the increasingly severe, interconnected and often irreversible impacts of climate change on ecosystems, biodiversity, and human systems; differing impacts across regions, sectors and communities; and how to best reduce adverse consequences for current and future generations.
>
> (IPCC 2022b)

Songwe et al. (2022) further warned that climate change is occurring at a faster pace than previously anticipated, the impacts and damage are greater than foreseen, and the time for remedial action is rapidly narrowing.

In the context of climate change, risk can arise from the dynamic interactions among climate-related *hazards*, the *exposure* and *vulnerability* of affected human and ecological systems.[1] Therefore, *mitigation*—taking strong actions to reduce emissions—must be viewed as an investment, a cost incurred now and in the coming few decades to avoid the risks of severe consequences in the future (Stern 2006). If these investments are made wisely, the costs of climate change will be manageable, and there will be a wide range of opportunities for growth and development along the way (IPCC 2022b).

Meanwhile, *adaptation* plays a key role in reducing *exposure* and *vulnerability* to climate change. IPCC's Sixth Assessment Report (2022b) has made it clear that

> with increasing warming, climate zones are projected to further shift poleward in the middle and high latitudes. In high-latitude regions, warming is projected to increase disturbance in boreal forests, including drought, wildfire, and pest outbreaks. In tropical regions, under medium and high GHG emissions scenarios, warming is projected to result in the emergence of unprecedented climatic conditions by the mid to late 21st century.

As such, we must heed these events and the damages they may cause. Adaptation is the process of adjusting actual or expected climate and its effects to moderate harm or take advantage of beneficial opportunities. In human systems, adaptation is often organized around *resilience*—bouncing back and returning to a previous state after a disturbance (IPCC 2022a). More broadly, the term describes not just the ability to maintain essential function, identity, and structure but also the capacity for transformation.

As mentioned, progress has been made in assessing the socioeconomic impacts of forest disturbances as well as responding to them. For instance, Holmes et al. (2008) is a nice synthesis of the early work done mostly by the US Forest Service, and Zhai and Ning (2022) is a latest review of the relevant literature. But many of the economic studies have failed to capture the ecological impacts of forest disturbances simultaneously, and vice versa. Thus, integrated economic and ecological assessments are called for and, whenever possible, an assessment should reflect the climate change damages in a spatially explicit manner. Also, damage assessments must be carried out for not just disturbances that have occurred in the past but also those that will occur in the future. Further, it is important to adopt adaptive management strategies suited to the local conditions, in response to potentially more intense events of climate change (IPCC 2020).

Below, let me use a recent study, conducted by Hsiang et al. (2017), to illustrate these points. As summarized by Pizer (2017), the four key components to value the benefits of reducing climate change include: projections of population, economic activity, and emissions; climate models to predict how small perturbations to baseline emissions affect the climate; damage models to translate climate change into impacts measured against the baseline economic activity and population; and a discounting scheme to convert the future damages of current incremental emissions into an appropriate damage value today. Together, these components can be used for computing the social cost of carbon (SCC)—an estimate of the economic cost (i.e., damages in monetary unit) resulting from emitting one additional ton of carbon dioxide ($CO_2$) into the atmosphere; conversely, it represents the benefit to society of reducing $CO_2$ emissions by one ton (Stern 2006, Nordhaus 2021). Partly determined by the chosen discounting scheme, the SCC is commonly used as a benchmark against which proposed or realized carbon prices or other mitigation initiatives are compared (Aldy et al. 2021, Carbon Pricing Leadership Coalition 2017).

Hsiang et al. (2017) developed a unified architecture to compute potential economic damages from climate change, and their risk-centered approach was based on longitudinal analyses of nonlinear, sector-specific impacts, supplemented with detailed energy system, inundation, and cyclone models. When applied to the US economy, they provided a probabilistic and empirically derived "damage function," linking global mean surface temperature to market and non-market costs using micro-level data. It was found that 3°C of warming would lead to a loss of ~2% of gross domestic product (GDP); 6°C of warming would lead to a ~6% GDP loss. By improving the damage models for the US, the study enhanced the credibility for future benefit estimates (Pizer 2017). The resultant analysis of the costs and benefits of climate change mitigation can form a cornerstone for more meaningful and durable policy.

While it may not be necessary for scholars of forest economics, policy, and finance to conduct this type of highly aggregate assessment, Hsiang et al. (2017) offer some highly valuable lessons regarding how similar studies can be done in the forest sector. These lessons include the ways to integrate various components of the economic and ecological systems, the downscaling of projected effects of climate change, and the adoption of probability-weighted outcomes based on carefully calibrated empirical data. By assuming a 3% constant discount rate, nonetheless, Hsiang et al. (2017) did not consider the ramifications of alternative discount schemes.

Another significant and complementary study, done by Gollier and Hammitt (2014), is a synthesis of the literature on the long-run discount rate controversy. They noted that the choice of the rate at which one should discount the long-term benefits of mitigating climate change is highly controversial.

> Both the level and the slope of the term structure of discount rates have been discussed intensively in relation to the determination of the social cost of carbon. Although some of the parameters of the problem are ethical and outside the scope of economic analysis, we claim that there are converging and convincing arguments in favor of using an annual real

risk-free discount rate going from approximately 4% to approximately 1% for maturities going from zero to infinity.

Gollier and Hammitt (2014) emphasized that investing in climate change mitigation yields highly uncertain future benefits, which should be considered in the selection of the discount rate.

As a latest effort following up with Hsiang et al. (2017) and Gollier and Hammitt (2014), Rennert et al. (2021) explored social discounting in a stochastic world in the context of deriving the SCC. Notably, Professor Stern (2006) already voiced that "a single constant discount rate would generally be unacceptable for dealing with the long-run, global, non-marginal impacts of climate change." He also stressed that even in a static world, there is no presumption that it is constant over time, as it depends on the way in which consumption grows over time. Using a high, constant discount rate decreases the assessed benefit of actions designed to reduce GHG emissions. So, the *Stern Review* did not employ a single discount rate but applied a stochastic approach whereby the discount rate varied with the expected outcomes, as discussed below, to reflect the interaction between consumption growth and the elasticity of marginal utility, in line with the model of Ramsey (1928). Accounting for risk in the stochastic world, of course, means that the expected mean or certainty equivalent discount rate will be below the discount rate for the mean expected outcome—a greater weight is applied to worse case outcomes, as per the insurance market (Dietz and Stern 2008, Gollier et al. 2008).

## Social discounting and cost of carbon

### Basics of social discounting

In fact, scholars have long realized that a constant, deterministic discount rate becomes increasingly problematic for such long-horizon challenges as global climate change, loss of biodiversity, and thinning of stratospheric ozone (e.g., Weitzman 1998, Gollier 2002). One justification for a modified discounting approach stems from uncertainty in the discount rate. Weitzman (1998) showed that if one is uncertain about the future trajectory of (risk-free) discount rates, and uncertain shocks to the discount rate are persistent, the *certainty-equivalent* discount rate declines with the time horizon toward the lowest possible level. Weitzman (2001) went on to propose an operational framework to resolve the discount-rate dilemma, centered on the notion of "gamma discounting." His numerical example suggested using the following stepwise approximation of within-period marginal discount rates for long-term public projects: Immediate Future about 4% per annum; Near Future about 3%; Medium Future about 2%; Distant Future about 1%; and Far-Distant Future about 0%. As a matter of fact, this kind of social discounting scheme has been implemented in the UK and France, among others (Gollier and Hammitt 2014).

To understand the seemingly perplexing nature of social discounting, it is worthwhile to go back the fundamental economics briefly. Economists view GHG emissions and the resulting climate change as the canonical example of a negative externality, as the cost it entails (or the benefits of emission reduction and removal) spill outside of the market worldwide and are not directly reflected in prices (Stern 2006, Nordhaus 2021). That is, for activities associated with this type of broad externality, the costs, benefits, and prices are not properly aligned, leading to the common phenomenon of free riding. But this externality can be addressed through Pigouvian taxation or other means. In particular, the basic economic theory suggests that an optimal tax on $CO_2$ emissions (or carbon tax) be set equal to the SCC, for which incremental damages are measured along an optimal emissions trajectory (Pigou 1920, Nordhaus 2021).

As the primary economic measure of the costs and benefits of mitigating climate change, the SCC is typically estimated using an integrated assessment model (IAM), such as the DICE (Dynamic Integrated Climate-Economy) model built by Nordhaus (1982) and an improved version of it exemplified by Hsiang et al. (2017). However, many of the factors underlying the SCC are uncertain. As such, the need for more robust climate policy implies that we should update the SCC over time to refine central estimates and the range of uncertainty as our scientific understanding progresses, which is exactly what Rennert et al. (2021) did in response to the report by the National Academies of Sciences, Engineering, and Medicine (NASEM 2017) that had recommended the creation of an integrated framework underlying the SCC calculation.

Rennert et al. (2021) first exposited that when inputs to a module are uncertain (e.g., due to uncertainty about the climate's response to emissions or about future economic growth), modelers may incorporate that uncertainty through Monte Carlo analyses by taking draws of probability distributions of each random variable. The result is a distribution of SCCs, often summarized by its expected value. For example, the federal government's interim estimate of $51/ton $CO_2$ for 2021 (U.S. Interagency Working Group, or IWP, 2021) reflects the expected value of the SCC over uncertainty in the climate's warming response and scenarios of economic growth and population at a 3% constant discount rate. The requirements of SCC estimation, however, pose significant hurdles for generating such projections. One is the time horizon: Given the long-lived nature of GHGs in the atmosphere, the SCC needs to account for discounted damages 200 to 300 years into the future (NASEM 2017). This time horizon makes the discount rate a major factor for the SCC. The SCC estimate for 2021—$51/ton with a 3% discount rate—would be about $121/ton at a 2% discount rate (Rennert et al. 2021). That 1 percentage-point difference alone would more than double the SCC and, by implication, greatly strengthen the economic rationale for substantial emissions reductions sooner. To fully appreciate this logic of social discounting, though, we need to revisit the Ramsey formula.

## The Ramsey formula

Stern (2006) used a social discount rate based on the "Ramsey formula," which includes a term for inherent discounting, also called the pure rate of time preference:

$$s_t = \gamma + \eta\, g_t$$

where $s_t$ is the social discount rate, $\gamma$ the pure time preference rate, $\eta$ the marginal elasticity of utility (or the curvature of an iso-elastic utility function), and $g_t$ the rate of growth of per-capita consumption. Here, time subscripts are used to refer to the compound average value of the indexed variable from today (time 0) to year $t$.

One justification for discounting is the concept of declining marginal utility of consumption. Intuitively, a $100 cost in a future in which society has grown dramatically wealthier should be valued less, from today's perspective, than the same $100 cost in a relatively poor future with stagnant growth. This finding is embodied by the classic equation derived in Ramsey (1928) that relates the consumption discount rate ($s_t$) to the rate of consumption growth ($g_t$) over time. If average consumption growth to year $t$, $g_t$, is uncertain, as it is in the probabilistic socioeconomic scenarios discussed earlier, then the discount rate in year $t$, $s_t$, is also uncertain, which gives rise to a stochastic discount factor. A stochastic discount rate resulting in a declining certainty-equivalent risk-free rate stems from a straightforward application of Jensen's inequality to a stochastic discount factor (Rennert et al. 2021). At the same time, if the payoffs to investments in emission reduction and removal (ER&R) are correlated with future income, the effective *risk-adjusted* rate could be higher when the correlation is positive or lower when it is negative (Gollier and Hammitt 2014). Of course, the actual sign and magnitude of this correlation are subject to empirical analysis (van den Bremer and van der Ploeg 2021).

## Implications of a declining discount rate

Incorporating the full range of economic uncertainty in the SCC underscores the importance of adopting a stochastic discounting approach to account for uncertainty in a unified manner. By addressing the challenges of projecting long-term economic growth, population, and greenhouse gas emissions, Rennert et al. (2021) were able to calibrate discounting parameters for consistency with those projections. This has made improvements on alternative approaches, such as non-probabilistic scenarios and constant discounting that have been used by governments but do not fully characterize the uncertainty distribution of fully probabilistic model input data or corresponding SCC estimate outputs. Using a stochastic discount factor will discount damages most in states of the world where they accrue to rich future generations and correspondingly discount them least in states where the future is poor. Importantly, executing the stochastic discount rate alongside stochastic damages via the above equation explicitly captures risk aversion and the correlation between the discount rate and climate damages, which implies that no *ex post* risk adjustment to the discount rate is necessary.

In short, uncertainty about future consumption may be addressed either through adjustments to the discount rate or by replacing uncertain flows of consumption with certainty-equivalent ones. Following Weitzman (1998,) Stern (2006) adopted the latter approach, and Rennert et al. (2021) presented an updated empirical case for it. Even though we cannot claim that the debate over social discounting has been fully settled, it has become clearer that such a scheme, alongside the centrality of an SCC, is now well established. The effective discount rate declines with the passage of time because increasingly greater relative present-value weight is being placed on the states with the lower rates. In comparison, the higher-discount-rate states are relatively less important over time because their present value has been more drastically shrunk by the power of compound discounting at the higher rates (Gollier and Hammitt 2014, Rennert et al. 2021). Nevertheless, much remains to be deliberated as to how we reconcile the different perspectives of and approaches to social discounting and finance climate mitigation actions.

## Financing under the term structure of discount rates

While the social discounting scheme, as described above, has gained broad support from economists, disagreement and debate have persisted. Among the critics of the *Stern Review* are Professor Nordhaus and his associates. Nordhaus (2007) argued against the use of a low and declining discount rate, asserting that "The *Review*'s unambiguous conclusions about the need for extreme immediate action will not survive the substitution of assumptions that are more consistent with today's marketplace real interest rates and savings rates." But later, Nordhaus (2021) wrote in a more consolatory tone: there are two ways of using the Ramsey equation as a framework for discounting in global warming or other long-run questions. One is the *prescriptive view*, in which analysts argue for the ethically just goods discount rate. This is the approach taken in Cline (1992) and Stern (2006). This approach generally yields a low goods discount rate, around 1% per year. A second approach is the *descriptive approach*, advocated by Lind et al. (2013) and Nordhaus (1994), which is used in various versions of the DICE model. This approach assumes that investments to slow climate change must compete with investments in other areas. The benchmark should therefore reflect the opportunity cost of investment. The descriptive approach yields a market rate of return in the neighborhood of 5% per year when risks are appropriately included.

Viewed from the perspective of "a term structure of discount rates," however, we see that the existence of different approaches to discounting is not necessarily contradictory. Indeed, it can be well understood if we contextualize the different investments in terms of their intrinsic characteristics, including domains, missions, and maturities. Gollier and Hammitt (2014) explained that social and market discount rates translate our collective values toward the future into key economic variables. Societies with a low discount rate value the future more than do societies with a higher discount rate, and the way we discount long-term benefits expresses our responsibilities toward future generations. Common wisdom suggests that our market economy tends to be short sighted; it does not sufficiently value the long-term impacts of our actions— rarely looking beyond 30 years within the commercial domain. This indicates that people often use discount rates that are too high compared with what would be desirable from the viewpoint of intergenerational welfare.

Investing for the future implies a transfer of consumption from current generations to future generations. In a growing economy, this process means transferring consumption from the poor to the wealthy. If we recognize a collective ethical concern in favor of reducing inequalities, the discount rate can be seen as the minimum rate of return of the investment that is required to compensate for the increased intertemporal inequality that the investment generates. The larger our aversion to inequality, the higher the discount rate. This is the essence of the Ramsey (1928) rule. Moreover, historic evidence shows that market forces induced the private sector to use a rate of between 1% and 2% to discount risk-free projects during the past century (Gollier and Hammitt 2014).

Therefore, it is not all that hard to come to the consensus that the market can provide private goods, in which case the investment horizon is short, and a high and constant market discount rate can be and has been applied. On the other hand, mitigating climate change will provide a global public good over an extremely long time; it is hardly realistic to expect that the private sector will automatically step up to make the much-needed investments. To ensure the net-zero transition, the state has a crucial role to play by restructuring its policies, such that it not only sets the legal framework, builds the relevant infrastructure, and funds the basic research but also makes direct investments along with subsidizing investments to be made by the private sector (Coyle 2022). Therefore, in making direct and indirect investments to mitigate climate change, it is reasonable and acceptable for the government to adopt an appropriate social discounting scheme.

To be sure, market-based solutions, like emission trading and carbon taxing, should be sought, so long as the government can make necessary changes to the underlying institutional arrangements and regulatory system. Similarly, the government can introduce franchising, bidding, and other market-oriented means to attract private-sector investments in climate mitigation programs and projects. This is the distinction between the provision and production of a public good that Professor Ostrom (2010) elucidated. In addition, the government must mobilize the private sector, including business and other non-governmental organizations, through their social responsibility in general and multinational companies' net-zero commitments in particular, as discussed in the proceeding chapter. Even with these steps being taken, unfortunately, the pace of market creation and private sector participation in solving the climate crisis have been slow and limited (Meng and Simpson 2022, Songwe et al. 2022).

In their report to the COP 26 and 27 Presidencies, *Finance for climate action: Scaling up investment for climate and development*, Songwe et al. (2022) argued that acting on climate is about transforming our economies, particularly our energy systems, through investing in net zero, adaptation, resilience, and natural capital. Achieving this transformation requires strong investment and innovation and the right scale of finance of the right kind and at the right time. They estimated that the world needs a breakthrough and a new roadmap on climate finance that can mobilize $1 trillion per year in external finance that will be needed by 2030 for emerging markets and developing countries other than China. There is a significant role for public policy and government action to foster investment, and complementary roles for the private sector, multilateral development banks, international financial institutions, and concessional finance of various forms. Note that internal finance of developed countries is not included here.

Moreover, Songwe et al. (2022) asserted that powerful multipliers can emerge from the complementary strengths of all sources of finance. An overall financing strategy must be built, based on a granular understanding of the different areas being financed. It would utilize the complementary strengths of different pools of finance to ensure the right scale and kind of finance for different needs. It would blend finance with different costs and maturities, including grants and concessional finance, to match the financial returns and risks from investments. It would therefore consider different strands of finance as a complementary package, rather than simply focusing on the aggregate number.

## Time value of physical carbon

Interestingly, a few studies have called for an inclusion of the time value of physical carbon sequestered and stored by forest ecosystems (e.g., Richards 1997, van Kooten 2018, Parisa et al. 2022). Parisa et al. (2022) claimed that "the full potential of natural systems to store carbon has not been leveraged because policymakers have required long-term contracts to compensate for permanence concerns, and these long-term contracts substantially raise costs and limit deployment." Therefore, they contended that the preference for early action leads them to conclude that multiple tons of short-term storage of carbon in ecosystem stocks could be considered to have equal value—as measured by the social cost of carbon—as 1 ton of carbon sequestered permanently. This equivalence could be used to quantify the value of short-term carbon storage, thereby removing one of the most significant barriers to participation in the carbon market and enabling the full climate mitigation potential of the land sector to be realized.

Specifically, Parisa et al. (2022) began by noting that a critical factor that had slowed the implementation of land-based options as carbon offsets has been the concern about the stability of forest carbon stocks—permanence. Because forest ecosystems are susceptible to natural and human disturbances that could cause some or all of the stored carbon to be emitted subsequently, many analysts have been skeptical about the durability, and hence the value, of forest carbon offsets (Thamo and Pannell 2016, Wagner 2022). Typically, crediting rules require forest-based offsets to ensure that any project activities, or carbon used to offset emissions, are maintained and managed on the site "permanently," often taken to mean up to 100 years. Also, suppliers typically have to carry insurance and place a large proportion of the potential credits into a buffer pool that cannot be sold (Zhao et al. 2023). Parisa et al. (2022) thus asserted that the approaches currently used to manage permanence in the carbon market are not based on economics, and as a result, raise costs, reduce participation, and lower the supply of potential credits. A change that could help the private market flourish is agreement on the role of short-term carbon offsets and an effective and valid approach to quantify and value short-term carbon storage in ecosystems.

To circumvent this problem, a static horizon ton-year approach has actually been suggested as a way to account for physical tons of carbon stored for a short period of time so that they can be traded with $CO_2$ emitted through energy combustion (Korhonen et al. 2002, Chomitz and Lecocq 2003). But Parisa et al. (2022) proclaimed that the static horizon concept relies on an arbitrary end date for the comparison of two different carbon flux streams. Further, this approach relies on a comparison of undiscounted carbon fluxes over time, which weighs future storage of carbon the same as current storage of carbon, even though economic models rely on discounting to appropriately weight welfare outcomes over time. So, Parisa et al. (2022) emphasized that for the question of permanence in land-based sinks, failing to discount the future benefits and costs of land-based carbon streams versus energy emissions could lead to less-than-optimal investments in forests or other land-based sinks. Carbon accounting frameworks that use ton-years without properly valuing carbon over time, or other approaches that ignore the time-value of money, could systematically overestimate costs and thus tilt investment decisions away from nature-based solutions.

Economic approaches that utilize carbon rental concepts avoid this problem by valuing short-term carbon storage through carbon rental (Marland et al. 2001, Sohngen and Mendelsohn 2003), which values a ton of carbon stored for a year with a carbon rent derived from the market price, or the social cost of carbon, appropriately discounted. Renting short-term storage in forests or other nature-based solutions requires energy emission sources to hold an equivalent stock of rented carbon tons indefinitely to net-out their $CO_2$ emission. However, it is also possible to exploit the relationship in value between a current emission and a delayed emission to derive a formula that expresses the number of tons ($N$) that need to be held for 1 year (or "$n$" years) to equilibrate the present value of 1 ton of current emissions with $N$ tons of delayed emissions. Parisa et al. (2022), using a standard model of the global carbon cycle, showed how multiple tons of short-term storage of carbon in ecosystem stocks have the same economic value as 1 ton of carbon sequestered permanently. The resulting formulation, they believe, can be used in a carbon trading market to allow participation by individual landowners of forests who intend to hold their carbon stocks only for short periods.

I hope that such a "formula can be used as the basis for market trading so all tons stored or emitted can be traded in current years while accounting for permanence." Furthermore, I wish that as a consequence of adopting the formula, it is likely to "increase market participation and lower the transaction costs of trading between sources of emissions and individual units of land that can generate offset credits." However, I wonder whether the solution is the only one to the problem and if we need additional work on institutional arrangements to that end. There are a few matters for us to consider and elaborate on more carefully. First, we know that the market prices of carbon have been generally low—less than US $10 per ton of $CO_2$ (Midkiff et al. 2022), much lower than the estimated SCC of US$51 per ton. Given the low carbon prices, its annual rental value must be very low, which may not result in the expected participation and thus impact. Therefore, it is more compelling to raise the price of forest carbon credits closer to the SCC, which may do much greater good than taking care of the carbon rental value. Also, it appears that the SCC should have captured the value of delayed emission or accelerated sequestration, as I discussed above. Moreover, emission reduction and removal should be treated equally. The urgency as well as importance of both reduction and removal of GHG emissions should already be reflected in the updated SCC through a social discount scheme. That is why we have not heard anyone arguing for a similar treatment of GHG emissions.

Also, considering permanence will not necessarily impair participation if the institutional barriers are removed. As I have discussed elsewhere (Chapters 2 and 4), small landholders with a few tracts of trees can participate in a carbon sequestration/storage program for offsetting credits if their lands are bundled up by certain aggregators. Of course, these aggregators can use a formula like what Parisa et al. (2022) advocated to attract landowners' participation on a flexible time basis. It should be pointed out, though, that while such intermediaries exist, their current practices of low purchase price, credit reservation, and heavy fees are all disincentives hampering landowners' participation (DeFries et al. 2022). The government must do its part to remove these and other institutional hurdles in facilitating the participation of anyone, including small landowners. In short, the problem identified by Parisa et al. (2022) can be well addressed through aggregating or pooling under the jurisdictional approach to climate governance (Zhao et al. 2023).

## Concluding remarks

The primary objectives of this chapter were to identify and elaborate how to address the risk and uncertainty that the forest sector faces in a world of intense climate change and to elucidate the rationale of and approach to social discounting entailed in public finance for climate actions, including NbSs. Along the way, I also made my case on whether and how a time value of physical carbon might be considered when analyzing forest sector NbSs. Hopefully, these efforts will provide a clearer orientation and more effective deployment of appropriate methods and tools to tackle the risk and uncertainty, a more conducive framework for the benefit and cost analysis of public projects and programs, and a stronger justification for taking more substantive and urgent actions by the government, as well as the private sector.

In summary, as far as climate change is concerned, risk and uncertainty are more about the potential damages and thus human society's insufficient commitments to the reduction and removal of GHG emissions. Integrated economic and ecological assessments are needed for the forest sector, and both *ex post* and *ex ante* damage assessments should be pursued (IPCC 2022a). It is also important to formulate and adopt adaptive management strategies suited to the local conditions in response to possibly more intense climate events (IPCC 2022b). Meanwhile, coping with damages of climate change over a long horizon requires us to confront the challenge of social discounting in assessing the costs and benefits of its mitigation.

Regarding governments' investments in climate mitigation, the arguments are converging in favor of using a social discounting scheme—starting with a modest rate and letting it decline gradually as the time horizon becomes longer, which is vastly different from a high and constant rate commonly used in commercial discounting over a relatively short horizon (Gollier and Hammitt 2014). But the existence of different approaches to discounting is not necessarily contradictory; rather, it can be well reconciled if we examine different investments in terms of their intrinsic characteristics, including domains, missions, and maturities. Anyway, acting on climate is about transforming our economies, particularly our energy systems, and achieving this goal requires strong investment and innovation. In addition to direct jurisdictional commitments to climate financing, there is a significant role for public policy and government action to foster investment and innovation while complementing roles for the private sector, multilateral development banks, international financial institutions, and many others (Songwe et al. 2022).

Finally, my deliberations in this chapter have additional policy implications to the forest sector climate actions. First, it is noteworthy that like many individuals (Thaler and Sunstein 2009), forest landowners tend to be risk averse in their decision making (Hyde 2012). Coupled with their frequent lack of upfront funds for investing in afforestation, reforestation or forest management (Amacher et al. 2009), this calls for businesses, especially large corporations who are committed to ESG (environmental, sustainability, and governance) goals, including net zeros, to provide them with the needed payments at the time of investment in anticipation of the carbon offset credits they will be able to generate at a reasonable price in the near future (FAO 2016). Doing so will allow many small landowners to defray the expenses of stand establishment (via afforestation or reforestation) and treatment (via thinning and other management practices), which will go a long way in increasing not just their engagement but also their effective participation in carbon offsetting as partially reflected in the improved forest growth and structural conditions, in addition to the generation of other co-benefits to ecosystems and human wellbeing.

Meanwhile, we can now appreciate and approach government forest ownership and management more thoughtfully. According to the latest Global Forest Resource Assessment report (FAO 2020), as much as 73% of the world's forests are government owned and managed, and a large part of them are used for biodiversity conservation and soil and water protection. Governments are also urged to substantially increase investments in restoring degraded forests and other ecosystems (UN 2020). Thus, public forests have assumed a major role and are expected to assume an even greater role in the provision of various ecosystem services (ESs) of public good nature. Of course, some ESs, like soil and water protection, are local or regional public goods, while others, including ER&R and biodiversity conservation, are national or global public goods. Coupled with the potential damage of climate change, certainly, this calls for proper social discounting in assessing the costs and benefits of its mitigation using a social discounting scheme. Furthermore, it necessitates a more prudent consideration of the question of devolved management of public forests. While it is necessary to devolve the management of some public forests to communities and other local organizations to improve the incentives and outcomes of management (Yin et al. 2016), caution is warranted to maintain the continuity and stability of public forest management.

## Note

1 See the IPCC report, *Impacts, Adaptation and Vulnerability* for details, as well as definitions, of these concepts (www.ipcc.ch/report/ar6/wg2/).

## References

Aldy, J.E., Kotchen, M.J., Stavins, R.N., Stock, J.H. 2021. Keep climate policy focused on the social cost of carbon. *Science* 373 (6557): 850–852. doi:10.1126/science.abi7813.

Amacher, G.S., Ollikainen, M., Koskela, E. 2009. *Economics of Forest Resources*. Cambridge: MIT Press.

Carbon Pricing Leadership Coalition. 2017. *Report of the High-Level Commission on Carbon Prices*. World Bank Group.

Chomitz, K.M., Lecocq, F. 2003. *Temporary Sequestration Credits: An Instrument for Carbon Bears*. World Bank Policy Research Working Paper 3181.

Cline, W.R. 1992. *The Economics of Climate Change*. Washington, DC: Institute for International Economics.

Coyle, D. 2022. The double transformations. *Project Syndicate* (December 9).

DeFries, R., Ahuja, R., Friedman, J., Gordon, D.R., Hamburg, S.P., Kerr, S., Mwangi, J., Nouwen, C., Pandit, N. 2022. Land management can contribute to net zero. *Science* 376 (6598): 1163–1165.

Dietz, S., and N. Stern. 2008. Why economic analysis supports strong action on climate change: A response to the Stern Review's critics. *Review of Environmental Economics and Policy* 2 (1): 94–113.

Food and Agriculture Organization of the United Nations (FAO). 2016. *Forestry for a Low-Carbon Future: Integrating Forests and Wood Products in Climate Change Strategies* (FAO Forestry Paper 177). Rome, Italy.

Food and Agriculture Organization of the United Nations (FAO). 2020. *Global Forest Resource Assessment Report*. Rome, Italy.

Gollier, C. 2002. Discounting an uncertain future. *J. Public Econ.* 85:149–166.

Gollier, C., Koundouri, P., Pantelides, T., Boone, J., Geoffard, P. 2008. Declining discount rates: Economic justifications and implications for long-run policy. *Economic Policy* 23 (56): 757–795.

Gollier, C., Hammitt, J.K. 2014. The long-run discount rate controversy. *Annual Review of Resource Economics* 6 (1): 273–295.

Griscom B.W., Adams J., Ellis P.W., Houghton R.A., Lomax G., Miteva D.A., Schlesinger W.H., Shoch D., Siikamäki J.V., Smith P., et al. 2017. Natural climate solutions. *PNAS* 114 (44): 11645–11650.

Henderson, J.D., Abt, R.C., Abt, K.L., Baker, J., Sheffield, R. 2022. Impacts of hurricanes on forest markets and economic welfare: The case of Hurricane Michael. *Forest Policy and Economics* 140 (102735).

Holmes, T.P., Prestemon, J.P., & Abt, K.L. 2008. *The Economics of Forest Disturbances: Wildfires, Storms, and Invasive Species* (Vol. 79). Springer Science & Business Media.

Hsiang, S., Kopp, R., Jina, A., Rising, J., Delgado, M., Mohan, S., Rasmussen, D.J., Muir-Wood, R., Wilson, P., Oppenheimer, M., Larsen, K., Houser, T. 2017. Estimating economic damage from climate change in the United States. *Science* 356: 1362–1369.

Hyde, W.F. 2012. *The Global Economics of Forestry*, 1st ed. New York: RFF Press/Routledge.

International Panel on Climate Change (IPCC). 2020. *Climate Change and Land: Summary for Policymakers*. Available at: www.ipcc.ch/srccl/.

International Panel on Climate Change (IPCC Working Group II). 2022a. *Climate Change 2022: Impacts, Adaptation and Vulnerability* (Summary for Policymakers). Available at: www.ipcc.ch/report/ar6/wg2/.

International Panel on Climate Change (IPCC). 2022b. *The Sixth Assessment Report*. Available at: www.ipcc.ch/assessment-report/ar6/.

U.S. Interagency Working Group on Social Cost of Greenhouse Gases (IWP). 2021. *Technical Support Document: Social Cost of Carbon, Methane, and Nitrous Oxide Interim Estimates under Executive Order 1399*. Washington, DC.

Korhonen, R., Pingoud, K., Savolainen, I., Matthews, R., 2002. The role of carbon sequestration and the tonne-year approach in fulfilling the objective of climate convention. *Environmental Science and Policy* 5: 429–441.

Lind, R.C., Arrow, K.J., Corey, G.R., Dasgupta, P., Sen, A.K., Stauffer, T., Stiglitz, J. E., Stockfisch, J.A., Wilson, R. 2013. *Discounting for Time and Risk in Energy Policy.* Washington, DC: RFF Press.

Marland, G., Fruit, K., Sedjo, R. 2001. Accounting for sequestered carbon: The question of permanence. *Environmental Science and Policy* 4: 259–268.

Mendelsohn, R. 1994. Property rights and tropical deforestation. *Oxford Economic Papers* 46: 750–756.

Meng, B., Simpson, A. 2022. To activate hope, activate capital. *Project Syndicate* (December 19).

Midkiff, D., Yin, R.S., Zhang, H. 2022. *Carbon Finance and Funding for Forest Sector Climate Solutions: A Synthesis of Principles, Mechanisms, and Developments* (MSU Forestry working paper).

National Academies of Sciences, Engineering, and Medicine (NASEM). 2017. *Valuing Climate Damages: Updating Estimation of the Social Cost of Carbon Dioxide.* Available at: www.nap.edu/catalog/24651/valuing-climate-damages-updatingestimation-of-the-social-cost-of.

Ning, Z., Sun, C.Y. 2019. Carbon sequestration and biofuel production on forestland under three stochastic prices. *Forest Policy and Economics* 109 (102018).

Nordhaus, W.D. 1982. How fast should we graze the global commons? *American Economic Review* 72 (2): 242–246.

Nordhaus, W.D. 1994. *Managing the Global Commons: The Economics of Climate Change.* Cambridge, MA: MIT Press.

Nordhaus, W.D. 2007. A review of the Stern Review on the economics of climate change . *Journal of Economic Literature* 45 (3): 686–702.

Nordhaus, W.D. 2014. Estimates of the social cost of carbon: Concepts and results from the DICE-2013R model and alternative approaches. *Journal of the Association of Environmental and Resource Economists* 1 (1-2): 273–312.

Nordhaus, W.D. 2019. Climate change: The ultimate challenge for economics. *American Economic Review* 109: 1991–2004. Available at: https://doi.org/10.1257/aer.109.6.1991.

Nordhaus, W.D. 2021. *The Spirit of Green: The Collisions and Contagions in a Crowed World.* Princeton, New Jersey: Princeton University Press.

Ostrom, E. 2010. Beyond markets and states: polycentric governance of complex economic systems. *American Economic Review* 100: 641–672.

Parisa, Z., Sohngen, B., Marland, E., Marland, G, Jenkins, J. 2022. The time value of carbon storage. *Forest Policy and Economics* 144 (102840).

Patriquin, M.N., Wellstead, A.M., White, W.A. 2007. Beetles, trees, and people: Regional economic impact sensitivity and policy considerations related to the mountain pine beetle infestation in British Columbia, Canada. *Forest Policy and Economics* 9 (8): 938–946.

Pigou, Arthur C. 1920. *The Economics of Welfare.* Macmillan.

Pizer, W.A. 2017. What's the damage from climate change? *Science* 356: 1330–1331. https://www.palgrave.com/gp/book/9780230249318.

Ramsey, F.P. 1928. A mathematical theory of saving. *Economic Journal* 38 (152): 543–559. https://doi.org/10.2307/2224098.

Rennert, K., Prest, B.C., Pizer, W.A., Newell, R.G., Anthoff, D., Kingdon, C., Rennels, L., Cooke, R., Raftery, A.F., Ševčíková, H., Errickson, F. 2021. *The Social Cost of Carbon: Advances in Long-Term Probabilistic Projections of Population, GDP, Emissions, and Discount Rates* (Brookings Papers on Economic Activity).

Resources for the Future (RFF) and New York State Energy Research and Development Authority (NYSERDA). 2021. *Estimating the Value of Carbon: Two Approaches.* Available at: www.rff.org/publications/reports/estimating-the-value-of-carbon-two-approaches/.

Richards, K.R. 1997. The time value of carbon in bottom-up studies. *Critical Review of Environmental Science and Technology* 27: 279–292.

Roe, S., Streck, C., Obersteiner, M., Frank, S., Griscom, B., Drouet, L., et al. 2019. Contribution of the land sector to a 1.5°C world. *Nature Climate Change* 9: 817–828. doi:10.1038/s41558-019-0591-9.

Sohngen, B., Mendelsohn, R. 2003. An optimal control model of forest carbon sequestration. *American Journal of Agricultural Economics* 85: 448–457.

Songwe, V., Stern, N., Bhattacharya, A. (2022) *Finance for Climate Action: Scaling up Investment for Climate and Development.* London: Grantham Research Institute on Climate Change and the Environment, London School of Economics and Political Science.

Stern, N. 2006. *Stern Review: The Economics of Climate Change.* Available at: http://mudancasclimaticas.cptec.inpe.br/~rmclima/pdfs/destaques/sternreview_report_complete.pdf).

Thaler, R.H., Sunstein, C. 2009 (updated edition). *Nudge: Improving Decisions About Health, Wealth, and Happiness.* New York: Penguin.

Thamo, T., Pannell, D.J. 2016. Challenges in developing effective policy for soil carbon sequestration: Perspectives on additionality, leakage, and permanence. *Climate Policy* 16: 973–992.

Van den Bremer, T.S., van der Ploeg, F. 2021. The risk-adjusted carbon price. *American Economic Review* 111 (9): 2782–2810.

Van Kooten, G.C. 2018. The challenge of mitigating climate change through forestry activities: What are the rules of the game? *Ecological Economics* 146: 35–43.

Wagner, G. 2022. Who pays for climate change? *Project Syndicate* (November 16).

Wang, Y., Li, L., Yin, R., 2021. A primer on forest carbon policy and economics under the Paris Agreement. *Forest Policy and Economics* 132 (102595).

Wang, B.W., Yin, R.S., Wen, Y.L., Zhang, H. 2023. Understanding and undertaking jurisdictional approaches to climate governance. *Environmental Research: Climate* (under review).

Weitzman, M.L. 1998. Why the far-distant future should be discounted at its lowest possible rate. *Journal of Environmental Economics and Management* 36 (3): 201–208. https://doi.org/10.1006/jeem.1998.1052.

Weitzman, M.L. 2001. Gamma discounting. *American Economic Review* 91 (1): 260–271.

Yin, R.S., Newman, D.H. 1996. The effect of catastrophic risk on forest investment decisions. *Journal of Environmental Economics and Management* 31 (2): 186–197.

Yin, R.S., Newman, D.H. 1997. Long run timber supply and the economics of timber production. *Forest Science* 43: 113–120.

Yin, R.S., Zulu, L., Qi, J.G., Freudenberger, M., Sommerville, M. 2016. Empirical linkages between devolved tenure systems and forest conditions: Primary evidence. *Forest Policy and Economics* 79: 277–285.

Zhai, J., Ning, Z. 2022. Models for the economic impacts of forest disturbances: A systematic review. *Land* 11 (9): 1608. https://doi.org/10.3390/land11091608.

# 9 Evaluating the evaluated impacts of forest carbon programs and projects

## Introduction

Economic research has become more empirical (Hamermesh 2013, Abadie 2021), and much of this shift has been associated with the widespread applications of impact evaluation (IE) methods for measuring the effects of programs or projects on outcomes of interest (Athey and Imbens 2017, Angrist and Pischke 2017). Consequently, evidence-based policymaking has been taken as a mantra of the day (Khandker et al. 2010, Ferraro and Pattanayak 2006).

As detailed below, forest policy analysts have been not only aware of this mega, enduring trend in economics, but also active in advancing their research agendas by taking advantage of it. Among other achievements, they have produced several high-quality and high-impact papers. The primary objective of this chapter is to review and critique some of the more recent studies on the impacts of REDD+ and other forest sector climate actions. Particular attention will be devoted to deliberating how forest policy analysts can further improve the relevance and quality of their future work, including how they may contextualize their work more appropriately considering the provisions of the Paris Agreement and appropriate national (or subnational) circumstances.

There exists a long list of practical cases for which we can apply IE methods in measuring the impacts of forest policies. To name just a few, they include the establishment of a protected areas (PAs), the restoration of a degraded ecosystem, the adoption of a REDD+ program, the implementation of a biofuel initiative, and the devolution of forest tenure (Lan and Yin 2017). The IE methods, such as difference in differences (DID) and propensity score matching (PSM), have been widely applied to forestry contexts.

For instance, Andam et al. (2008) stated that conventional methods of evaluating the effectiveness of PAs could be biased because protection is not randomly assigned and protection might induce deforestation spillovers to neighboring forests. By adopting both PSM and DID to evaluate the impact on deforestation of Costa Rica's PA system during 1960–1997, they showed that protection reduced deforestation—~ 10% of the PAs would have been

DOI: 10.4324/9781003436652-10

deforested had they not been protected. In comparison, conventional approaches overestimated avoided deforestation by over 65%. Meanwhile, deforestation spillovers to unprotected forests were negligible.

With similar analytic techniques, Nolte et al. (2013) and Pfaff et al. (2014) confirmed that all protection regimes helped reduce deforestation in the Brazilian Amazon. For any given level of deforestation pressure, strict PAs avoided more deforestation than the so-called sustainable use areas. Indigenous lands were very effective at avoiding deforestation in locations with high deforestation pressure. In addition, IE methods have been used in assessing the effects of tenure devolution (Lawry et al. 2017, Holland et al. 2014), ecological restoration (Uchida et al. 2009, Liu et al. 2010), community forestry (Rahut et al. 2015), and REDD+ piloting (Sills et al. 2017).

Partly inspired by the earlier research advances and partly responding to the urgent call for more robust climate policy, forest resource economists have recently devoted a lot of effort to evaluating REDD+ projects and programs. Notable examples include Jayachandran et al. (2017), which evaluated a popular type of intervention—financial incentives for forest owners to keep their forests intact—in western Uganda, based on a randomized controlled experiment (RCE); and Roopsind et al. (2019), West et al. (2020), and Groom et al. (2022) that are observational studies of the impacts of REDD+ projects and programs in Guyana, Brazil, and Indonesia.

I will spend a large part of this chapter synthesizing these empirical cases and commenting on their merits and areas of improvement, here it suffices to say that these and other studies have convincingly demonstrated the great potentials of using IE methods in forest policy research. With appropriate framings and approaches, forest policy analysts can provide robust evidence of policy impact, enabling decision makers to better understand the interlinkages between human and natural systems in determining the policy outcomes and thus use this knowledge to effectively guide their actions in mitigating climate change and protecting, restoring, and managing forest ecosystems.

Of course, even if a cutting-edge IE method is deployed to assess the impact(s) of a particular policy, proper research design, including sampling procedure and covariate coverage, is essential. So, an overview of the fundamental design and methodologic issues of IE is in order before I proceed to presenting the case studies and deliberating the entailed policy and technical questions. This chapter is based on a short Research Trends commentary my colleague and I published in *Forest Policy and Economics* (Lan and Yin 2017). In writing this chapter, I have made extensive revisions and additions to it. The chapter is organized as follows: first, I overview the core IE concepts and methods, as well as sampling and data concerns in the next two sections; then, I present the selected case studies, before making comments on them and suggesting ways forward. A few closing remarks will follow in the final section.

## The essential methodology

### Basic concepts

Athey and Imbens (2017) concisely summarized:

> The gold standard for drawing inferences about the effect of a policy is a randomized controlled experiment. However, in many cases, experiments remain difficult to implement, for financial, political, or ethical reasons, or because the population of interest is too small.... Thus, a large share of the empirical work in economics about policy questions relies on observational data—that is, data where policies were determined in a way other than through random assignment.
>
> (p. 3)

Obviously, if we can execute randomized controlled experiments (RCEs), it will then be straightforward to determine the targeted impact of a project or program—the difference between the outcome of the treatment group and that of the control group. Note that "treatment" is often used interchangeably with terms like "intervention," "manipulation," and "event," and comparisons are made between the treatment group and control group (Abadie 2021, Khandker et al. 2010). In case RCEs are financially expensive and/or ethically or politically infeasible, however, an IE exercise must be conducted based on observational data. If done well, it can be very informative and insightful. But IE work has faced challenges. Included in them are: first, how we can properly identify a counterfactual—what would have happened to the treated group of individual units, or participants, if a policy had not been adopted; second, how we can remove or at least control the confounding factor(s) that may induce correlation between the policy under evaluation and the outcome that is not indicative of what would happen if the policy could be adopted (Imbens and Rubin 2015).

Unfortunately, what would have happened to the treated group if a policy had not been adopted is unobservable (Athey and Imbens 2017). It is thus natural to deal with this so-called "missing data" problem by creating a reasonable control group that has not been treated to determine the policy impact on the treated group (Khandker et al. 2010); those who received treatment would have had outcomes like those in the comparison group in the absence of treatment. So, it is presumed that only one of the two potential outcomes, $Y_i$, is observed for each sample unit $i$ in the basic formulation of an IE model. That is:

$$Y_i = Y_i(W_i) = \begin{cases} Y_i(0) & if\ W_i = 0 \\ Y_i(1) & if\ W_i = 1 \end{cases}$$

where $W_i \in \{0,1\}$ indicates the treatment received. Then, the average treatment effect for the treated, or *ATT*, is

$$ATT = \frac{1}{N} \sum_{i|W_i=1} \{Y_i(1) - Y_i(0)\}$$

where $N = \sum_i W_i$ the number of treated units, for which we only observe $Y_i(1)$. Dealing with this problem has led to several techniques for measuring the counterfactual effect, $Y_i(0)$, to determine the ATT (Imbens and Rubin 2015).

Understandably, each of these techniques carries its own assumptions about the nature of potential selection bias in policy targeting and participation, and these assumptions are crucial to developing the appropriate model in determining causal effects (Khandker et al. 2010). Selection bias is the bias introduced by the selection of individuals, groups, or data for analysis in such a way that proper randomization is not achieved, with the result that the sample obtained is not representative of the population intended for analysis (Angrist and Pischke 2008).

## Main analytic methods

The commonly used modeling techniques include difference in differences, propensity score matching, panel data models, and regression discontinuity.

*Difference in Differences* (or DID) is an often-used IE method in which outcomes are observed for both the treated group ($W_i = 1$) and the control group ($W_i = 0$) for a pre-treatment period $t_1$ and a post-treatment period $t_2$. So,

$$y_{it} = \beta_0 + \beta_1 d2 + \beta_2 dT + \beta_3 d2dT + u_{it}$$

where $y_{it}$ is outcome of interest, $dT = 1$ for treatment group ($W_i = 1$) and 0 otherwise, $d2 = 1$ for the post-treatment period ($t_2$) and 0 otherwise, $d2dT$ is the interaction term between $d2$ and $dT$, $u_{it}$ is random error, and $\beta_1, \beta_2, \beta_3$ are parameters to be estimated. Specifically, $\beta_0$ is the average outcome for the control group prior to the treatment ($t_1$), $\beta_1$ captures changes in $y$ caused by aggregate factors even in the absence of the treatment, and $\beta_2$ captures possible differences in outcomes between treatment and control groups prior to the treatment ($t_1$). The coefficient for the $ATT$ is $\beta_3$, which is defined as

$$\widehat{\beta_3} = (\overline{y_{i2}}(1) - \overline{y_{i1}}(1)) - (\overline{y_{i2}}(0) - \overline{y_{i1}}(0))$$

where $\overline{y_{i2}}(1)$ and $\overline{y_{i1}}(1)$ are the mean outcome for the treatment group in period $t_2$ and $t_1$, and $\overline{y_{i2}}(0)$ and $\overline{y_{i1}}(0)$ are the corresponding values for the control group.

In addition to clear-cut cases of treatment vs. control groups, we encounter situations of different treatment types or intensities. The above DID estimator remains applicable in the latter case so long as a proper indicator of the treatment types/intensities can be formulated and the confounding factors considered (Finkelstein 2007). One key assumption of the DID estimator is that the average outcomes for the treatment and control groups would follow parallel trends in the absence of the treatment (Khandker et al. 2010). If it does not hold, then the ATT estimate will be biased. Thus, alternative methods have been developed (Heckman et al. 1997).

*Propensity Score Matching* (PSM) constructs a statistically determined comparison group based on a model of the probability of participating in the treatment, using observed characteristics. Participants are matched to nonparticipants based on this probability or propensity score. The ATT is then calculated as the mean difference in outcomes across the two groups. In addition to DID, there are several numerical methods to determine the mean difference (Khandker et al. 2010). The validity of PSM is based on: (i) unobserved factors do not affect participation; and (ii) there is sizable common support in propensity scores across the two groups (Imbens and Rubin 2015). Note that although PSM can be deployed even if the analyst has only cross-sectional data—observations at a single point of time and use of a simple matching method other than PSM—this type of data and method, especially in a small sample, will make it less likely to identify a truly meaningful counterfactual and thus a reliable estimate of the policy impact.

*Panel Data Models* (PDMs) can be employed to explore the advantages of panel data by running the following model:

$$y_{it} = \lambda_t + \tau w_{it} + \gamma x_{it} + c_i + u_{it}, t = 1, 2, \ldots, T$$

where $w_{it} = 1$ if unit $i$ is treated at time $t$ and 0 otherwise, $x_{it}$ is a vector of covariates, $c_i$ is an unobserved individual-level effect, $\lambda_t$ accounts for aggregate time effects, and $u_{it}$ are the idiosyncratic errors (Wooldridge, 2009). Coefficient $\tau$ is the average treatment effect. This model can be estimated with a fixed effect or first differencing method, provided that the treatment indicator $w_{it}$ is exogenous, conditional on unobserved heterogeneity (Wooldridge 2009). If $w_{it}$ is correlated with unit-specific trends, a correlated random trend model can be used. The availability of panel data also allows for selection bias on unobserved characteristics to vary with time. Here, selection bias on unobserved characteristics can be corrected by using an instrumental variable (IV) method to find a variable (or instrument) that is correlated with participation but not so with unobserved characteristics affecting the outcome (Wooldridge, 2009). This instrument is then used to predict participation. Again, a proper indicator of the types and intensities of participation can be used as a treatment proxy in this setting, which allows the analyst to deal with issues like lag time and spatiotemporal correlations as well.

*Regression Discontinuity* (RD) methods exploit discontinuities in policy rules (e.g., eligibility requirements) and thus incentives or ability to receive a discrete treatment. Participants and nonparticipants are compared in a close neighborhood around the eligibility cutoff. The boundary demarcation of a protected area and the public assistance to local community members below the official poverty line are just two of the many familiar examples. To model the effect of a particular policy on individual outcomes $y_i$ through an RD method, one needs a variable $S_i$ that determines participation eligibility with an eligibility cutoff of $s_*$. The estimating equation is $y_i = \beta S_i + \varepsilon_i$, where individuals with $s_i \leq s_*$, for example, receive the treatment, and individuals with $s_i > s_*$ are not eligible to participate. Individuals in a narrow band above and below $s_*$ need to be "comparable" in that they would be expected to achieve similar outcomes prior to the treatment.

## Sampling and data considerations

Here, my deliberations fall into the following two categories: the first category is concerned with the technical strategies of sampling and data generation, while the second category is about how to respond to data constraints where limited observations are available because only a small number of large units, often at the aggregate level, are affected by the treatment.

### *Strategies of sampling and data generation*

Normally, data for an IE study should come from at least two periods of time, featuring times of before and after the policy treatment if it is a discrete and quick action (Khandker et al. 2010). Otherwise, the data may cover periods of early and late treatment, which seems more common given that many policies or programs take years, if not decades, to be fully implemented. Therefore, it must be clearly articulated whether the outcome of interest is a short-term or long-term effect (Lu and Yin 2020). Additionally, a careful description of the underlying situation, or contextualization, will go a long way in presenting the case under consideration, justifying the sampling strategy, specifying the empirical model, and thus achieving a less biased and more robust coefficient estimation (Lu and Yin 2020).

The next question is what data and variables are needed in estimating the outcome of interest. There are several points to be made here. First, keep in mind the relevance of variables. In several studies of China's Sloping Land Conversion Program—a large ecological restoration initiative—for instance, household and land characteristics are commonly covered, whereas the broader policy settings and social-ecological conditions are ignored, leading to potentially biased estimates of the impacts (Yin et al. 2014). Certain variable(s) can be captured only if attention moves beyond the narrow scope of sample units—whether individuals, households, or other local and more aggregate organizations (Yin 2009). Having data beyond this narrow scope will help better identify the confounding factors and/or capture the key covariates.

Moreover, if possible, variables should be directly measured. Some analysts have the tendency of making a quick run in their survey and data generation by constructing discrete dummy (yes/no) and categorical proxy variables (e.g., low, average, high levels of income). This is a superficial approach, which makes it difficult to derive convincing and substantive empirical results. So, subjective ranking or perception should be avoided whenever possible. Of course, if the situation warrants only subjective ranking or perception, the survey should be properly designed and executed. In this situation, having some prior, preliminary field and/or literature investigation and community or focus group interviews may become even more essential (Yin et al. 2016).

### Synthetic control as a response to data constraints

In the last few years, synthetic controls (SCs) have been applied to study the effects of many policy issues, which evaded rigorous analysis previously (Abadie 2021). SC methods were originally proposed in Abadie and Gardeazabal (2003) and Abadie et al. (2010), aiming to estimate the effects of aggregate interventions, that is, interventions that are implemented at an aggregate level impacting a small number of large units (such as cities, regions, or countries), on some aggregate outcome of interest. Traditional regressions require large samples and many observed instances of the intervention of interest and, consequently, they are often ill-suited to estimate the effects of infrequent events on aggregate units. On the other hand, the use of time-series techniques to estimate medium and long-term effects of an intervention is complicated by the presence of shocks to the outcome of interest.

More specifically, SC is based on the idea that, when the units of observation are a small number of aggregate entities, a combination of unaffected units often offer a more appropriate comparison than any single unaffected unit. Various data-driven techniques for variable selection can be adopted to evaluate the predictive power of alternative sets of predictors. In essence, the procedure divides the pre-intervention periods into an initial training period and a subsequent validation period, with SC weights to be computed using data from the training period only. The validation period can then be used to evaluate the predictive power of the resulting SC. The SC methodology formalizes the selection of the comparison units using a data-driven procedure, with a SC defined as a weighted average of the units in the donor pool (Abadie 2021). Under appropriate conditions, SCs provide substantial advantages as a research design method in the social sciences.

## Case studies

### Randomized controlled experiment

To put REDD+ into action, one needs to identify policy interventions that effectively reduce deforestation and forest degradation. Based on a creative REC design, Jayachandran et al. (2017) evaluated a popular type of intervention, namely, financial incentives for forest owners to keep their forests intact—a payment for ecosystem services (PES) program—in western Uganda. Designed and implemented by a local conservation nonprofit,

Chimpanzee Sanctuary and Wildlife Conservation Trust (CSWCT), the program offered private forest owners (PFOs) in 121 villages payments if they refrained from clearing trees. Sixty villages were randomly selected to be in the treatment group. To participate, PFOs had to enroll all their primary forest. During the two-year program from 2011 to 2013, enrollees who complied with the contract received a cash payment of $28/ha in 2012 US dollars per year. CSWCT hired monitors who conducted spot checks of enrollees' land to verify whether any recent tree-clearing occurred. The program also provided additional payments in exchange for planting tree seedlings.

The research team measured the impact of the program on forest cover by analyzing satellite imagery in a novel manner. They tasked a high-resolution commercial satellite, QuickBird, to take images of the study area at baseline and endline and classified each pixel as tree-covered or not using eCognition, an object-based image analysis software product. With a multispectral resolution of 2.4m and a panchromatic resolution of 0.6m, the QuickBird pixel size is smaller than the crown of a typical mature tree. Next, they assessed how many additional hectares of tree cover the program caused by comparing PFOs' land in treatment and control villages. The high-resolution data also detected additional loss of trees due to selective tree-cutting within the forest, i.e., forest degradation. They further estimated program impacts on secondary outcomes collected via a household survey and analyzed cost-effectiveness by calculating the amount of $CO_2$ emissions delayed by the program and its monetary value.

They found that in the control group, the average tree loss per village between baseline and endline was 9.1% of baseline tree cover, which was higher than most estimates of Uganda's national rate of forest loss based on changes in the perimeter of the forest. Then, they estimated that the program caused an increase in tree cover, relative to the control group, of 5.55ha per village. The treatment group experienced net tree loss but considerably less than the control group, and the treatment effect is statistically significant at the 5% level. The average tree loss in the treatment group was 4.2%—about half of the loss in the control group. Put differently, the program averted 0.326ha of deforestation per eligible PFO, with an averted $CO_2$ of 183.5 tons. The average payment per PFO was $37.80 over the two years. Thus, the program paid $0.20 to delay each ton of $CO_2$ emissions. Their estimate of total program costs—incentive payments to forest owners plus program administration costs—was only $0.46 per averted ton of $CO_2$.

By randomly assigning who would be eligible for the program, this study improved on the existing ones, which were subject to concerns about bias in the estimates due to self-selection into the program or targeting by program administrators based on unobservables (Imbens and Wooldridge 2009). The study is also commendable for using satellite images with high resolution, enabling the authors to detect selective tree-cutting in addition to clearcutting. In many parts of the world, thinning of forests has led to significant forest loss (Kissinger et al. 2012). Moreover, their analysis showed the program's cost effectiveness in comparison to other options for reducing carbon emissions. Nonetheless, the RCE was costly and technically demanding; but the team was able to overcome these challenges to produce a rare RCE study of broad interest. At the same time, its short duration made it impossible to investigate the longer-term effect, let alone its permanence.

## Observational studies

### Guyana

Norway and Guyana entered one of the first bilateral REDD+ programs in 2009, with Norway offering to pay $250 million to Guyana if annual deforestation rates remained below 0.056% from 2010 to 2015. To quantify the impact of this national REDD+ program, Roopsind et al. (2019) constructed a counterfactual trajectory of annual tree cover loss using synthetic matching, allowing them to quantify tree cover loss that would have occurred in the absence of the program. As I overviewed above, SC-based matching is an empirical method that constructs a counterfactual time-series scenario, in this case "what tree cover loss would have occurred without the REDD+ program?" to evaluate the causal impact of the policy intervention. The authors claimed that this approach explicitly accounts for unobserved drivers of the outcome (i.e., factors that influence the outcome that we do not know of or do not have data for).

It was found that the program reduced tree cover loss by 35% during the implementation period, equivalent to 12.8 million tons of avoided $CO_2$ emissions—strong evidence that the program met the additionality criterion of carbon accounting and thus achieving the intended impact of national REDD+. The cost of the avoided $CO_2$ emissions was estimated to be $19.53/ton. Interestingly, the authors also found that tree cover loss increased after the payments ended; therefore, without continued payments, forest protection is not guaranteed.

This REDD+ program was implemented at the national jurisdictional level and employed a combined incentive-crediting baseline for emissions reductions. As the authors remarked, results such as these based off rigorous impact evaluation of climate mitigation programs can help countries refine their climate action plans post-2020. One of the direct policy implications is that the application of a sliding scale deforestation reference baseline, with increasing penalties as deforestation rises, seems to have empowered both the buyer and provider of carbon credits.

### Indonesia

During 2000–2016, about half of Indonesia's annual emissions were generated from deforestation, forest degradation, peatland decomposition, and peat fires (Groom et al. 2022). The country's partnership with Norway, again, included a pledge of $1 billion to fund results-based REDD+ payments. Central to this partnership was a moratorium on granting new concession licenses by district governments for the conversion of primary dryland and peatland forests into new palm oil, timber, and logging concessions. Initiated in 2011, the moratorium covered 69 million ha of forest—most of Indonesia's forest estate. Additional restrictions on the conversion of peatland forest, affecting all concession types, were carried out in 2017, and, in late 2018, a new three-year moratorium on palm oil concessions was also imposed.

Groom et al. (2022) argued that using a historical baseline as a counterfactual would provide weak evidence for the moratorium's effect, because deforestation in any given year would vary due to stochastic natural processes and economic factors. Thus, it is unlikely that observed deviations from the average deforestation rate can be meaningfully related to the moratorium performance. Their counterfactual mimicked what would have happened in those areas had the moratorium not been implemented. But moratorium areas were not randomly assigned; estimating its impact on deforestation was therefore hampered by preexisting differences in levels and the likely trajectory of deforestation between moratorium and non-moratorium areas. Also, district governments could continue to issue licenses for new concessions in forestland outside moratorium areas, and spatial spillovers is a common concern in forest conservation, confounding estimates of moratorium impact.

To address these issues and isolate the moratorium's impacts on deforestation, they adopted a matched triple difference strategy, which was applied to Global Forest Change data at the 1.2km × 1.2km scale between 2004 and 2018. Their outcome variable was "forest cover" in ha. Concessions established within the moratorium's 2011 boundaries prior to its start were allowed to continue operating, business as usual (BAU), after 2011. The moratorium was effective in protecting no more than 1 ha of dryland forest in each grid cell by 2018, or about 0.651%. Avoided dryland forest loss represented, at most, 0.03% of all land covered by the moratorium in 2012, rising to 0.22% by 2018. These estimates translated into cumulative emission reductions of 67.8 million tons (Mt) of $CO_2e$ to 86.9Mt $CO_2$ in 2011–2018. Further analysis indicated no leakage within a considerable distance from the moratorium.

They finally inferred that the average annual emissions reductions amounted to around 10.4 Mt $CO_2$ to 13.0Mt $CO_2$ (including peatland forest estimates and factoring in belowground carbon), comprising 0.38 to 0.47% of annual aggregate national emissions and 0.83 to 1.05% of emissions from the forest sector. Accordingly, the emissions avoided due to the moratorium were 10.3 to 12.9% of the 29% (unconditional) NDC target in 2020, and these shares fell to 7.8 to 9.8% when they were examined in relation to the 41% (conditional as well as unconditional) NDC target.

*Brazil*

West et al. (2020) adopted a similar technique to estimate the effects of 12 voluntary REDD+ projects in the Brazilian Amazon. The projects were implemented since 2008 and certified under the Verra Carbon Standards (VCS) before May 2019. The key question is whether these projects have generated additional impacts. To address that question, the authors compared the crediting baselines established ex-ante by the voluntary projects to counterfactuals they constructed ex-post. They also examined trends in forest loss in buffer zones around the project areas following project implementation to assess the plausibility that any reductions in deforestation might have been displaced.

The authors reported no significant evidence that the projects had mitigated forest loss. Deforestation was consistently lower in the project site than in the SC in only four of the projects, and this difference was outside the confidence interval around zero established by the placebo tests. Moreover, they discovered that the crediting baselines significantly overstate deforestation in comparison to the counterfactual estimates based on SCs. According to the projects' ex ante estimates, up to 24.8 million carbon offsets could have been generated by the REDD+ interventions by 2017, but the VCS database indicated that only 5.4 million tradable credits from these projects had been certified and made available to offset emissions from private and public sources by that year.[1]

Thus, no systematic evidence proved that the certified carbon offsets claimed by the projects (with one exception) were associated with additional reductions in deforestation in the REDD+ areas. They concluded that if carbon credits would be expected to reflect changes in emissions caused by REDD+, using historical baselines could lead to excess credits for projects when deforestation at the regional level drops below the historical baseline.

## Discussion

It seems more sensible to bring out the common features of the case studies first before examining the major challenges they encountered. So, my discussion unfolds around the data and methods first, and then along the principles and rules of carbon accounting and crediting.

### *Technical merits*

It is worth noting that all the case studies used geospatial cover data derived from satellite imagery. For purposes of IE, the use of geospatial data is reasonable and indeed advantageous, given their availability and manipulability. The use of high-resolution satellite images by Jayachandran et al. (2017) is particularly innovative, as it can generate timely and accurate, continuous observations that may not come easily from a national forest inventory (NFI) system given its focus on highly aggregate jurisdictional units and low frequency of iteration—typically every 5–10 years. The recent accessibility of a spatially explicit cadastral database allowed West et al. (2020) to define a donor pool of rural properties like the REDD+ project areas in their counterfactual identification.

Meanwhile, it should be made clear that if the spatial resolution is coarse (say, one pixel covers at least 1km × 1km), then the data may not have the requisite precision that a quality IE demands. For one thing, the accuracy of the land-use map derived from the Landsat imagery with a pixel size of roughly 30m × 30m rarely gets higher than 90% (Shi et al. 2017). Groom et al. (2022) and Roopsind et al. (2019) used the Global Forest Change data with a resolution of 1.2km × 1.2km. It remains to be seen whether and under what circumstances the geospatial data can be accepted by the UNFCCC and other international agencies, as well as the science community, as a substitute for forest cover information in carbon offsetting and crediting.

Because of using forest cover as the outcome variable, another issue is that the other carbon pools—litter and below ground biomass, as well as soil—could have been omitted or inaccurately estimated. Indeed, some calculations are needed to convert the estimated cover condition into above-ground biomass and carbon, which may introduce errors compared to stock volume measured by an NFI. Groom et al. (2022) mentioned that they factored in belowground carbon in their estimation; but it remains unclear whether they factored in only belowground biomass carbon or both belowground biomass and soil carbon pools were captured. Surely, an argument can be made that these carbon pools are simply not covered by the bilateral contracts. If so, however, a clear distinction should be made between the bilateral efforts to reducing deforestation and forest degradation and thus carbon emissions from these sources and the requirement of the UNFCCC for a Party to show how it has achieved its climate mitigation targets as specified in its NDC.

Another feature of all four studies is their shared emphasis on identifying counterfactuals to determine the impact of a REDD+ program or project. As already elaborated, it is straightforward for an RCE study to generate such a counterfactual because, by definition, the treatment is randomly assigned and thus enables an analyst to identify a comparison group and control for the confounding factors and possible endogeneity of participation for the treated group in estimating the policy effect. On the other hand, it is challenging for any observational study to identify a reliable counterfactual given the impracticality of assigning the treatment randomly and the existence of confounding factors (Imbens and Rubin 2015). Another added difficulty is that at the relevant aggregate level—national or subnational jurisdictions—it is hard to find a comparison group, leading to the need for creating some sort of SCs (Abadie 2021). But there are alternative pathways along this direction, causing the variation of outcomes assessed by alternative sets of predictors. Thus, care must be taken when such a study is done.

### *Framing challenges*

First, it is more meaningful to focus on national or subnational jurisdictional initiatives. At the project level, the heterogeity and transaction costs can be huge while the accountability and confidence tend to be low.[2] All of this makes it hard for the international community to reach down to individual projects. Rather, they should be integrated into a national or subnational jurisdictional initiative and administered accordingly. That is why I feel that using SCs is more compatible with a focus on national and subnational jurisdiction commitments, in line with the requirements of the Paris Agreement. A project orientation is not well aligned with the Agreement's predicament on the nationally determined contributions.

In addition, if a REDD+ country makes both conditional (with international payments) and unconditional (without international payments) commitments, like what Indonesia has done, a further question is whether the same baseline should apply to both situations. As I explored in Chapters 1 and 2, for an unconditional commitment, once the base year is chosen, the baseline for a country is the carbon emission or removal generated in that year. In contrast, the baseline for a conditional commitment is the counterfactual scenario (not a single point)—only if the outcome of the REDD+ program/project is above and beyond its corresponding BAU level can it be legitimately credited and rewarded.

Thus, the observational studies all alerted that crediting baselines based on the historical trends could cause overstatement of deforestation. Groom et al. (2022) observed that submitted to the United Nations Framework Convention on Climate Change in 2016, Indonesia's forest reference level (FRL) was based on the average annual deforestation rate between 1990 and 2012[3]; and Indonesia's FRL provided the basis for a proposed $103.8 million payment from the Green Climate Fund for an estimated 20.3 million tons of carbon dioxide equivalent (MtCO$_2$e) reduction in carbon emissions between 2014 and 2016. In 2017, Indonesia reportedly reduced emissions from deforestation and forest degradation by 11.2MtCO2e; subsequently, Norway announced that it would pay Indonesia $56.2 million based on a carbon price of US$5/t CO$_2$e. The authors showed that the counterfactual they formulated allowed to remove any remaining deviations in deforestation trends prior to the moratorium commencing in 2011.

West et al. (2020) were aware of that the REDD+ projects in Brazil were not part of the country's NDC; the voluntary carbon market was apparently poorly developed, with a lack of regularity. Realizing that the crediting baselines assumed higher deforestation than counterfactual forest loss in SC sites, they suggested an alignment of project- and national-level carbon accounting. That is, the REDD+ projects could become part of the national or subnational (jurisdictional) program of emission reduction, with a predefined baseline as well as default carbon-stock values. They went on to note that imposing one common baseline would in turn facilitate the inclusion of carbon emission reductions claimed by decentralized initiatives into national GHG emission inventories, ensuring consistency in the treatment of leakages, and avoiding double-counting reductions.

It is true that national and subnational baselines are typically based on historical data as well and, thus, are not any more likely to capture contemporaneous deforestation drivers and their dynamism, as pointed out by West et al. (2020). Even though there are limitations, however, historical data can be used for baseline development by national and subnational jurisdictions, especially if they make unconditional commitments to emission reductions from deforestation and forest degradation or from other sources. Furthermore, while we can question the accuracy of a baseline, it does not have to capture contemporaneous deforestation drivers and their dynamism. West et al. (2020) has exposed the major challenges encountered in voluntary carbon projects. Similarly, the study of Coffield et al. (2021) of climate benefits of California forest carbon offset projects disclosed the difficulties of robustly quantifying the additionality and thus the need for improving both governance design and implementation protocol.

Relatedly, credits from the voluntary projects are generally issued after a third-party audit (i.e., verification) every one to five years. However, while crucial, the third-party involvement in auditing and verification can be problematic and thus must be carefully monitored and, if necessary, better structured and regulated (Zadek 2023). In comparison, Roopsind et al. (2019) started their analysis with a focus on the national REDD+ program and a clear discussion of the principles of additionality, permanence, and no leakage in efforts of reducing tree cover loss and carbon emissions in Guyana. This seems more appropriate.

## Concluding remarks

This chapter was mainly motivated to synthesize and review some of the recent empirical studies on the impacts of REDD+ and other forest sector climate actions. To fully understand and appreciate these studies, I began the chapter with an overview of the general methodology and core technical issues entailed in impact evaluation. Following a summary of four case studies, I discussed their merits and areas of improvement, including how such an IE can be more adequately contextualized.

In summary, forest carbon accounting, credit taking, and progress tracking at the national and subnational levels must reflect the principles of additionality, permanence, and leakage avoidance measured against an established baseline. As such, a well-executed IE should reflect these principles in the context of particular jurisdiction(s) before any evidence is generated. While the existing literature has been aware of these principles and rules, they may not have been upheld clearly or consistently. One key issue is the selection of a baseline, which differs corresponding to the types of commitments made by a Party. If it is an unconditional commitment, then the baseline is emission and/or removal levels at the point of time defined in the NDC, in which case, historical data, insofar as they are accurate, can be directly used in determining the baseline and thus compliance with the principles. Otherwise, if it is a conditional commitment, then the baseline is the BAU scenario covering a period, again, specified in the NDC. While it is possible to decide that scenario based on historical data, there exist disagreements over how long the period of the data should cover, what estimation approach should be taken, and so on. Of course, answers to these questions have to do with the variability embedded in the historical data on the one hand and a jurisdiction's intention to manipulate the data and rules in its favor on the other.

Brazil is a good example to illustrate this point. As demonstrated by West et al. (2020), the country experienced substantial reductions in deforestation prior to the last a few years. Therefore, the Brazilian government could use the historical data covering an extended period and a baseline that is determined by a simple averaging to justify its continued, albeit moderated, reduction in deforestation in the compliant time and thus to claim larger amounts of carbon credits. But they countered that the BAU scenario should be set by considering the historical drops in deforestation rate and adopting an SC method that is less prone to incorrectly attribute changes in deforestation to REDD+. While I do not discount the relevance of the SC method, I wonder if it is realistic for the UNFCCC to adopt it and, even if it is, how the Conference of Parties can have a set of rules regarding the data coverage, predictor selection, and estimate conversion. Obviously, it will be a negotiated settlement, before which some piloting activities must be undertaken.

At the same time, we should not forget the fact that funding for REDD+ initiatives, mostly based on bilateral agreements with Norway and Germany, is limited in amount and poorly linked up with the UNFCCC or the principles of carbon accounting stipulated in the Paris Agreement. Moreover, a number of non-REDD+ countries that have been implementing forest sector climate actions have not received adequate research attention yet. Regardless of the circumstances, however, more efforts should also be devoted to the permanence conditionality of a program or project in the future. In this context, it is worthwhile to examine the related issues of climate governance, safeguards including forest tenure, and incentives and co-benefits of forest sector climate actions (IPCC 2020, FAO 2016).

Finally, applications of IE methods vary by their underlying assumptions regarding how to identify a counterfactual and how to resolve selection bias in estimating the treatment effect of a policy. RCEs involve a random assignment of the intervention across the sample; the statuses of treatment and control subjects exhibiting similar preprogram characteristics are then tracked over time (Khandker et al. 2010). RCEs have the advantage of avoiding selection bias at the level of randomization. In the absence of an RCE, the remaining option is to perform an observational study using PSM, DID, and other methods. A PSM method compares treatment effects across participant and matched nonparticipant units, with the matching being conducted on observed characteristics. It assumes that selection bias is based only on observed characteristics; thus, it cannot account for unobserved factors affecting participation.

In comparison, a DID method assumes that unobserved selection is present and that it is time invariant—the treatment effect is determined by taking the difference in outcomes across treatment and control units before and after the policy intervention. An IV method can be used with cross-section or panel data, and in the latter case it allows for selection bias on unobserved characteristics to vary with time. Using an IV method, we can correct selection bias on unobserved characteristics by finding a variable, or instrument, that is correlated with participation but not so with unobserved characteristics affecting the outcome; this instrument is then used to predict participation (Imbens and Rubin 2015).

An RD method is similar to the IV method as it introduces an exogenous variable that is highly correlated with participation (Angrist and Pischke 2008). If panel data are available, it becomes feasible to address such issues as fixed or random effects, autocorrelations, and time lags using alternative specifications. Finally, an SC method can be deployed to overcome data constraints associated with interventions that are implemented at an aggregate level affecting a small number of large units.

The soundness of an estimated impact depends on how justifiable the assumptions are on the comparability between treated and comparison groups as well as the exogeneity of policy targeting across treated and non-treated areas. It other words, any well-construed analysis should maintain internal and external validity (Angrist and Pischke 2008). Wider and more competent use of these methods and procedures will benefit forest policy analysts immensely, especially at a time when they are called upon to contribute to the design, implementation, and assessment of important projects and programs, including REDD+ and other forest sector climate actions. It is exciting to pursue more rigorous IE studies that draw up a quality research design, build datasets with broader coverage and better enumeration, and elucidate the complex relationships across temporal and spatial scales (Yin 2009, Abadie 2021).

## Notes

1 One unit of offsets or credits is normally 1 ton of $CO_2$.
2 An investigative report in *The Guardian* on 18 January 2023 revealed the effectiveness of REDD+ and forest carbon offsets was very low as the Peruvian projects lacked the proper oversight and monitoring standards.
3 Here, a minor distinction should be made: If forest cover is used as the outcome variable, then the corresponding analysis is based on the FRL, instead of the forest reference emission level (FREL).

## References

Abadie, A. 2021. Using synthetic controls: Feasibility, data requirements, and methodological aspects. *Journal of Economic Literature* 59 (2), 391–425. https://doi.org/10.1257/jel.20191450.

Abadie, A., Diamond, A., Hainmueller, J. 2010. Synthetic control methods for comparative case studies: Estimating the effect of California's tobacco control program. *Journal of the American Statistical Association* 105 (490): 493–505.

Abadie, A., Gardeazabal, J. 2003. The economic costs of conflict: A case study of the Basque Country. *American Economic Review* 93 (1): 113–132.

Andam, K.S., Ferraro, P.J., Pfaff, A., Sanchez-Azofeifa, G.A., Robalino, J.A. 2008. Measuring the effectiveness of protected area networks in reducing deforestation. *PNAS* 105 (42): 16089–16094.

Angrist, J.D., Pischke, J.S. 2017. Undergraduate econometrics instruction: Through our classes, darkly. *Journal of Economic Perspectives* 31 (2): 125–144.

Angrist, J.D., Pischke, J.S. 2008. *Mostly Harmless Econometrics: An Empiricist's Companion*. Princeton, NJ: Princeton University Press.

Athey, S., Imbens, G.W. 2017. The state of applied econometrics: Causality and policy evaluation. *Journal of Economic Perspectives* 31 (2): 3–32.

Coffield, S.R., Vo, C.D., Wang, J.A., Badgley, G. et al. 2022. Using remote sensing to quantify the additional climate benefits of California forest carbon offset projects. *Global Change Biology* 28:6789–6806. https://doi.org/10.1111/gcb.16380.

Food and Agriculture Organization of the United Nations (FAO). 2016. *Forestry for a Low-Carbon Future: Integrating Forests and Wood Products in Climate Change Strategies* (FAO Forestry Paper 177). Rome, Italy.

Ferraro, P.J., Pattanayak, S.K. 2006. Money for nothing? A call for empirical evaluation of biodiversity conservation investments. *PLoS Biology* 4 (4): e105.

Finkelstein, A. 2007. The aggregate effects of health insurance: Evidence from the introduction of medicare. *The Quarterly Journal of Economics* 122 (1): 1–37.

Groom, B., Palmer, C., Sileci, L. 2022. Carbon emissions reductions from Indonesia's moratorium on forest concessions are cost-effective yet contribute little to Paris pledges. *PNAS* 119 (5): e2102613119.

Hamermesh, D.S. 2013. Six decades of top economics publishing: Who and how? *Journal of Economic Literature* 51 (1): 162–172.

Heckman, J.J., Ichimura, H., Todd, P. 1997. Matching as an econometric evaluation estimator: Evidence from evaluating a job training program. *Review of Economic Studies* 64 (4): 605–654.

Holland, M.B., Koning, F.D., Morales, M., Naughton-Treves, L., Robinson, B., Suarez, L. 2014. Complex tenure and deforestation: implications for conservation incentives in the Ecuadorian Amazon. In: Naughton-Treves, L., Alex-Garcia, J., Baird, I., Turner, M., Wendland, K. (eds), Land tenure and forest carbon management (Special Section). *World Development* 55: 21–36.

Imbens, G.W., Rubin, D.B. 2015. *Causal Inference for Statistics, Social, and Biomedical Sciences*. New York: Cambridge University Press.

Imbens, G.W., Wooldridge, J.M. 2009. Recent developments in the econometrics of program evaluation. *Journal of Economic Literature* 47 (1): 5–86.

International Panel on Climate Change (IPCC). 2020. *Climate Change and Land: Summary for Policymakers*. Available at: www.ipcc.ch/srccl/.

Jayachandran, S., de Laat, J., Lambin, E.F., Stanton, C.Y., Audy, R., Thomas, N.E. 2017. Cash for carbon: A randomized trial of payments for ecosystem services to reduce deforestation. *Science* 357: 267–273.

Kissinger, G., Herold, M., De Sy, V. 2012. *Drivers of Deforestation and Forest Degradation: A Synthesis Report for REDD+ Policymakers*. Vancouver, Canada: Lexeme Consulting.

Khandker, S.R., Koolwal, G.B., Samad, H.A. 2010. *Handbook on Impact Evaluation: Quantitative Methods and Practices*. Washington, DC: The World Bank.

Lan, J., Yin, R.S. 2017. Policy impact evaluation: Future contributions from economics. *Forest Policy and Economics* 83: 142–145.

Lawry, S., Samii, C., Hall, R., Leopold, A., Hornby D., Mtero, F. 2017. The impact of land property rights interventions on investment and agricultural productivity in developing countries: A systematic review. *Journal of Development Effectiveness* 9: 61–68.

Liu, C., Lu, J.Z., Yin, R.S. 2010. An estimation of the effects of China's forestry programs on farmers' income. *Environmental Management* 45: 526–540.

Lu, G., Yin, R.S. 2020. Evaluating the evaluated socioeconomic impacts of China's Sloping Land Conversion Program. *Ecological Economics* 177 (106785).

Nolte, C., Agrawala, A., Silviusb, K.M., Soares-Filhoc, B.S. 2013. Governance regime and location influence avoided deforestation success of protected areas in the Brazilian Amazon. *PNAS* 110 (13): 4956–4961.

Pfaff, A., Robalino, J., Lima, E., Sandoval, C., Herrera, L.D. 2014. Governance, location and avoided deforestation from protected areas: Greater restrictions can have lower impact, due to differences in location. In: Naughton-Treves, L., Alex-Garcia, J., Baird, I., Turner, M., Wendland, K. (eds), Land tenure and forest carbon management (Special Section). *World Development* 55: 7–20.

Rahut, D.B., Ali, A., Behera, B., 2015. Household participation and effects of community forest management on income and poverty levels: Empirical evidence from Bhutan. *Forest Policy and Economics* 61: 20–29.

Roopsind, A., Sohngen, B., Brandt, J. 2019. Evidence that a national REDD+ program reduces tree cover loss and carbon emissions in a high forest cover, low deforestation country. *PNAS* 116 (49): 24492–24499.

Shi, M.Y., Yin, R.S., Lv, H.D. 2017. An empirical analysis of the driving forces of forest cover change in northeast China. *Forest Policy and Economics* 78: 200–209.

Sills, E.O., Sassi, C.D., Jagger, P., Lawlor, K., Miteva, D.A., Pattanayak, K.P., Sunderlin, W.D. 2017. Building the evidence base for REDD+: Study design and methods for evaluating the impacts of conservation interventions on local well-being. *Global Environmental Change* 43: 148–160.

Uchida, E., Rozelle, S., Xu, J.T. 2009. Conservation payments, liquidity constraints, and off-farm labor: Impacts of the Grain for Green Program on rural households. *American Journal of Agricultural Economics* 81, 247–264.

West, T.A.P., Borner, J., Sills, E.O., Kontoleon, A. 2020. Overstated carbon emission reductions from voluntary REDD+ projects in the Brazilian Amazon. *Proceedings of the National Academy of Sciences* 117, 39: 24188–24194.

Wooldridge, J.M. 2009. *Introductory Econometrics: A Modern Approach*, 4$^{th}$ edn. Cincinnati, OH: South-Western College Publishing.

Woodridge, J.M. 2010. *Econometric Analysis of Cross Section and Panel Data*, 2$^{nd}$ edn. Cambridge, MA: MIT Press.

Yin, R.S., 2009. *An Integrated Assessment of China's Ecological Restoration Programs*. Dordrecht, The Netherlands: Springer.

Yin, R.S., Liu, C., Zhao, M.J., Yao, S.B., Liu, H. 2014. The implementation and impacts of China's largest payment for ecosystem services program as revealed by longitudinal household data. *Land Use Policy* 40: 45–55.

Yin, R.S., Zulu, L., Qi, J.G., Freudenberger, M., Sommerville, M. 2016. Empirical linkages between devolved tenure systems and forest conditions: Challenges, findings, and recommendations. *Forest Policy and Economics* 73: 294–299.

Zadek, S. 2023. Trouble with carbon markets. *Project Syndicate* (March 1).

# 10 Summary and outlook

**Synopsis**

The forest sector can play a major role in mitigating climate change, and this role is based on the likely multitude and magnitude of actions to be undertaken within it. Encompassed in these actions are not just reducing $CO_2$ emissions from deforestation and forest degradation but also removing $CO_2$ emissions from the atmosphere through expanded forest coverage, improved forest management, and enhanced carbon stock (IPCC 2020). Globally, land-based activities, including agriculture, forestry, wetlands, and bioenergy, could sustainably contribute about 30% (15 billion tons) of the $CO_2$ equivalent per year required to limit the increase in the global temperature to 1.5°C above pre-industrial levels by year 2050 (Roe et al. 2019). Forest sector activities could themselves offer two-thirds of the cost-effective nature-based solutions necessary to hold warming to less than a 2.0°C increase (Griscom et al. 2017).

Nonetheless, the international science and policy communities have encountered a series of questions in envisioning, designing, implementing, and evaluating concrete actions in the forest sector because of the lack of an adequate understanding of the fundamental policy, economics, and finance. These questions include:

1 What are the essential principles and rules of carbon accounting and assessment?
2 Why do we need a jurisdictional approach to governing forest sector climate actions?
3 What are the core elements and modeling systems of forest carbon economic analyses?
4 How should we handle the local-level joint production of timber and carbon properly?
5 How can we assess the role of forest sector climate actions in national decarbonization?
6 What is the varied carbon leakage of wood product trade derived from alternative views?

DOI: 10.4324/9781003436652-11

168  Summary and outlook

7  How can we move forward with carbon finance and funding for forest sector actions?
8  What are the risk and uncertainty involved in climate change mitigation and adaptation?
9  Why should we adopt a social discount rate in analyzing public climate investment?
10  In what ways can we improve the quality of impact evaluations of carbon policies?

Therefore, I set out to tackle these questions in this book. I began with an elaboration of the relevant provisions of the Paris Agreement itself in Chapter 1 and a perspective on governing forest sector climate solutions in Chapter 2. Then, I conducted a literature overview of carbon accounting and assessment at different levels of aggregation in Chapter 3, providing a requisite orientation of the landscape of economic inquiry. Next, I investigated mitigation efforts at the local and national levels in Chapters 4–5, before examining three broader issues that extend beyond national boundaries in Chapters 6–8: international trade, funding and finance, and risk, uncertainty, and social discounting. Finally, I reviewed and synthesized impact evaluations of forest carbon policies in Chapter 9, focusing on the popular REDD+ projects and programs. Whenever possible, I included examples or case studies from different parts of the world.

In this closing chapter, I offer my summary observations and recommendations for further inquiry and renewed policy approaches to forest sector climate solutions.

## Elaboration on the Paris Agreement

In Chapter 1, I highlighted the key elements of the Paris Agreement (PA) and deliberated their implications for forest sector actions to build a consistent understanding of the processes, principles, and rules pertinent to climate change mitigation and adaptation. Among the themes discussed were:

- the makeup of and information need for the nationally determined contribution (NDC),
- the processes of progress tracking and stock taking,
- jurisdictional and other approaches to climate governance,
- the determination of anthropogenic greenhouse gas (GHG) removals,
- the principles and rules of carbon accounting, and
- the forest sector data requirements and gaps.

Corresponding to the targets specified in its NDC, a Party must select appropriate indicators and use proper assumptions and methodologies to track progress. Then, it becomes feasible to conduct carbon accounting, as well as stock taking, following such principles and rules as additionality, permanence, and leakage measured against an established baseline or reference level (UNFCCC 2020). But there are large gaps between the available information and what is needed for counting forest sector actions. This is because even if Parties have reliable data for forest stock volume, from which the amount of carbon sequestered in the biomass can be estimated, they may not have good information on carbon fluxes associated with the forest soil pool, urban trees and forests, or trees outside of forests elsewhere (FAO 2020a). Improved measurement and monitoring are crucial if they desire to include these carbon pools in their offsetting plans.

At the same time, it appears difficult to get the results-based payments scheme carried out at scale, because it takes time and effort to resolve the challenges in identifying the reference levels of REDD+ projects and programs and thus their carbon additionality on the one hand. On the other hand, it has been a struggle for advanced economies to deliver their promised international finance. Also, while active and rapidly evolving, voluntary markets, including transactions in the Land Use, Land-Use Change and Forestry (LULUCF) arena, represent a tiny portion of the existing, let alone the expected, size of world carbon markets (Ecosystem Marketplace 2021). Against this backdrop, however, forest sector actions have gone way beyond the scope and significance of the REDD+ initiative[1].

Unlike the project orientation of the Kyoto Protocol, the NDC architecture underlines a jurisdictional approach to international climate governance—making and executing pledges by administrative bodies of a national government, with the obligations of subordinate bodies being nested within the national pledge (von Essen and Lambin 2021, DeFries et al. 2022). Meanwhile, the UNFCCC tracks the progress that has been made by a national government and assesses to what extent the pledged commitments will collectively meet the PA's temperature targets. As such, Chapter 2 was devoted to examining the distinctions, strengths, and executions of alternative approaches to governing climate solutions.

Compared to project-based approaches, jurisdictional ones have certain advantages (Wunder et al. 2020). They provide a greater likelihood for countries to carry out necessary actions and assume their responsibilities. Also, they make it more practical for the carbon accounting principles to be materialized, and jurisdictions tend to have a stronger capability of measurement and monitoring, reporting, and verification (MRV) in carrying out their commitments. Adopting jurisdictional approaches, thus, could have made what we learned under the Kyoto Protocol or from the experience of voluntary markets no longer adequate.

Indeed, certain principles and practices pertinent to the governance of forest sector climate solutions have not been well understood. For example, while most analysts are aware of the principles of additionality, permanence, and leakage avoidance for the accounting and evaluation of climate solutions, the question remains whether these principles should be followed primarily at the national level or at the individual project level. From the global perspective, it seems more appropriate to examine how these principles are followed by national jurisdictions. It would be unimaginable for the international community to check and verify how these principles are followed at the project level, given the likely number of them to be carried out within each jurisdiction and the heterogeneity of local circumstances (West et al. 2020).

Of course, the underpinning of a jurisdictional approach to climate governance has to do with the nature of climate as a global public good or externality (Nordhaus 2019), which calls for a broad, effective, and substantive participation of the whole global community in stabilizing the climate and controlling the temperature rise. Accordingly, every social organization should do its part within a unified framework to save costs and enhance effectiveness of climate actions; market-based and non-market measures should be subject to the chosen approach(s) to climate governance, and different components of the system operate cohesively (Ostrom 2010).

Still, we should recognize the limitations of jurisdictional approaches, such as political turnover, limited public capacity, and the lack of broader support and incentives—all of which can hamper the kind of long-term, sustained attention forest sector initiatives require to succeed. Employing a jurisdictional approach may also face hurdles in places with rampant corruption and weak enforcement (DeFries et al. 2022). Another concern for all approaches is the need for fair and equitable investments and benefits sharing (Pachauri et al. 2022, FAO 2016). But it is unrealistic to expect that the market or jurisdiction can easily resolve centuries-old problems of insecure land rights and procedural injustice (DeFries et al. 2022). So, REDD+ and other forest sector actions must combine national coordination and policy cohesion with meaningful local involvement in implementation (Parrotta et al. 2022).

## Empirical analysis of forest sector actions

### The alternative economic frameworks

Emission reduction and removal (ER&R) in the forest sector are commonly analyzed for the purposes of carbon accounting and assessment, and there exist alternative techniques to conduct these tasks. Forest inventory data are directly used in carbon accounting for determining an entity's ER&R, whereas inventory data are often linked up with some kind of modeling system of timber, or wood products, markets and thus used indirectly for assessment. As such, the existing literature on forest carbon accounting and assessment can be classified into two strands—one featuring the direct inventory-based technique and the other the indirect inventory-based technique.

According to Ohrel (2019), included in the alternative techniques are "historic reference and extrapolations, ecological models, forest sector economic models, integrated assessment models, and meta-analysis." But I took a different perspective. That is, there exist two basic techniques—direct and indirect inventory-based. The former, the same as the "historic reference and extrapolations" of inventory data described by Ohrel (2019), should be easy to understand. It makes future projections of and evaluates policy effects on forest conditions using historical data of growth, yield, area, and so on.

Omitting economics may lead to biased ER&R estimates. In most parts of the world, however, a market model of timber, or wood products, is simply not available, whereas inventory data are (FAO 2020b). Given this, only the direct inventory-based technique is feasible. It can also be argued that after all because we deal with $CO_2$ (and other GHGs) in the context of climate change mitigation, not timber per se, a timber market modeling system has to work backward to the inventory status before deriving the carbon stock and/or flux associated with the stock volume, not to mention other pools like forest soil and storage by harvested wood products (HWPs).

For a market-based modeling system, the same as Ohrel's (2019) "forest sector economic models," forest inventory is included as a driver of the timber supply, whereas the demand for timber is derived from the demands for wood products (Sohngen and Mendelsohn 2003). In other words, the core of a market model is made up of a set of wood products supply and demand equations subject to the maximization of social welfare—a sum of consumer and producer surpluses, with a module of forest inventory dynamics linked with the market model. Once developed, such a system can be used to make simulations and projections of the inventory and other variables of concern, including carbon.

The so-called "ecological models" are often used to capture the inventory response to environmental changes, such as carbon fertilization, nitrogen deposition, and/or forest fire (Grassi et al. 2021), and they are members of the family of direct inventory models (Nabuurs et al. 2010). "Integrated assessment models," or IAMs, are typically modeling systems that integrate models of economy, energy, and/or ecology to assess potential policy outcomes and tradeoffs, allowing insights not possible in single sector models (Nordhaus 2021). "Meta-analysis" is commonly deployed to extract findings and other useful information from existing studies of possibly different techniques (Roe et al. 2019), which may include direct and indirect inventory-based models, and even IAMs.

Both direct and indirect inventory-based modeling techniques need to consider land-use and/or management technology changes in making future projections (Johnston and Radeloff 2019). Depending on the degree of sophistication, these components may be simply some assumed scenarios, or carefully formulated shared socioeconomic pathways (SSPs) (IPCC 2022a). Furthermore, different methods can lead to discrepancies between the gross change of forest carbon stock and the counterpart induced by anthropogenic forces because they may not capture the indirect effects resulting from increase in atmospheric $CO_2$ concentrations and other factors.

In fact, a large fraction of ER&R resulting from indirect human activities are considered 'natural' in IAMs but 'human-caused' in inventories (Grassi et al. 2021). As such, there has been a persistent gap between inventory-based

## 172  Summary and outlook

estimation and IAMs in the carbon cycle literature. By combining these indirect emissions and removals from dynamic global vegetation models with the direct emissions from the IAMs, Grassi et al. (2021) eliminated most of the discrepancies. For now, we can determine whether the forest biomass and carbon dynamics are driven by human actions and whether we can conduct carbon accounting and assessment based directly on the nation forest inventory (NFI) information and the delineation given therein.

In addition to highlighting the potential uses and merits of different techniques in this chapter, the selected case studies from Canada, Russia, and Europe, as well as the whole world, illuminated that deployed properly, either of them can generate powerful policy insights for forest management. While highly aggregate, market-based modeling efforts remain indispensable to carbon accounting and assessment, more direct inventory-based, jurisdiction-oriented attempts are also necessary. It is crucial for the forest sector modeling community to continually update models with the most recent information and validate their outcomes independently and as part of larger model comparison efforts.

### *Empirical studies*

#### *Micro-level carbon accounting*

As a follow-up step, I presented a micro-level study of carbon accounting and offset credit determination in Chapter 4, based on the pine plantation forests in the US South (Yin et al. 1998). Studying forest sector carbon sequestration and storage at the micro level has attracted attention from many scholars, but there remain deficiencies in how to incorporate them into an analysis properly (Wang et al. 2021). One of the deficiencies is reflected in the frequently used Hartman modification to the Faustmann model for determining the optimal rotation and land value of a timber stand (Hartman 1976, van Kooten et al. 1995), which is not well-suited for dealing with a continuous process underlying forest carbon accounting or assessment (Li et al. 2020).

Predicated on the adoption of such a discrete model, existing studies (e.g., Ekholm 2016) have failed to conform to the PA accounting principles, treat timber and carbon as joint products appropriately, and capture the differentiated potentials of carbon storage by HWPs. Therefore, it is imperative to establish a more satisfactory framework of forest carbon accounting and explore the outcomes of alternative approaches. To that end, I formulated a forest-level profit function that treats timber and carbon as joint products coherently and captures the PA accounting principles and the differentiated potentials of carbon storage in HWPs.

My empirical analysis demonstrated that the outcomes indeed deviate a lot, with those derived from the Hartman-Faustmann approach exaggerating the amount of carbon offset credits a landowner can take by a factor of at least 2.76. Also, while adding carbon to the valuation has limited impact on

*Summary and outlook* 173

extending the optimal rotation age, its effect on raising the net revenue can be substantial. I believe that these findings are of broad interest, given the great desire for individual landowners to participate in climate actions in all parts of the world (FAO 2016, IPCC 2020).

My analysis also illustrated that properly accounted for, forest carbon offsetting actions are important and credit worthy and, thus, deserve to be promoted and rewarded. But determining the rotation length must now attend to HWPs, because of their markedly different service lives and the disproportionality between carbon stored in sawnwood and that in paper/paperboard. It is necessary to explore how to increase the production of sawnwood and wood-based panels to extend the overall life of carbon stored in HWPs. Due to these variations, coupled with the permanence principle, nonetheless, the potential of forest sector nature-based solutions (NbSs) to climate change mitigation may not be as great as some analysts have claimed (e.g., Hoel et al. 2014).

Similarly, complying with the PA's requirements for carbon accounting does not imply that smallholders who have only a few stands of trees cannot participate in an emission offsetting and trading schemes or are not eligible for taking credits from their contributions. However, it does mean that there should be intermediary agencies who can aggregate the smallholders and bundle their individual properties, not only to make their participation feasible and trustworthy but also to reduce the impediments and costs entailed in MRV (FAO 2016). Innovations in the brokerage of forest carbon investment and trading are thus needed, which in turn calls for an improved regulatory system of the voluntary as well as compliance markets.

Future research should extend this type of work to other FESs of the world to understand their differentiated potential and cost competitiveness. Further, I did not consider the possibility of generating biofuels and/or bio-chemicals from woody materials. In view of their promises to replace fossil fuels and/or non-renewable materials (Jonsson et al. 2021), it is yet another important opportunity for future research. In these endeavors, though, more field experiments of carbon sequestration by different FESs and storage in HWPs are warranted, given that much of the current work on carbon accounting is based on the relatively crude Tier I parameters of the IPCC (2006).

*The role of China's forest sector in decarbonization*

The national potential of China's forest sector in its decarbonization was investigated in Chapter 5. A question of international interest is: How significant a role can China's forest sector play in achieving the country's goal of net-zero emissions? To answer this question, it is essential to project China's carbon sequestration by FESs and storage in HWPs in addition to its carbon emissions (IPCC 2006). Recent research advances (Duan et al. 2021, Piao et al. 2009, Jiang et al. 2016, Johnston and Radeloff 2019), coupled with the accumulated information in these areas, makes it feasible to quantify the role that China's forest sector has played in offsetting its carbon emissions so far and the role it will likely play in the future.

I found that the actual gain of forest stock volume was $5.063 \times 10^9 m^3$ during 2006–2020, leading to an offsetting ratio of 6.99%. When combining carbon sequestered by forest biomass and soil and stored in HWPs, these actions could offset 11.76% of China's cumulative carbon emissions during 2006–2020. Overall, China's FESs and HWPs could offset 14.72% of its cumulative carbon emissions during 2006–2060, even under conservative assumptions. Also, my calculation suggested that China's forest sector carbon sequestration and storage could more than offset its cumulative emissions during the 2050s, implying that the country would achieve carbon neutrality a decade earlier than what the government has envisioned. In addition, it indicated that the government has been repeatedly conservative in setting its target for forest stock volume.

Other than elucidating the significance of negative emission technologies as reflected in FES sequestration and HWP storage (Fuss et al. 2018), I argued that these actions are more cost effectiveness in achieving China's ER&R targets. Duan et al. (2021) revealed that the models they reviewed gave China a social cost of carbon in 2050 between \$315 and \$2,240 per $tCO_2$ (in 2010 US dollars). In contrast, a recent study in northeast China suggests that the carbon costs of timber plantations are in the range of \$23.4 to \$45.1 per $tCO_2$ (Li et al. 2020).

Furthermore, investing in thinning and other stand treatments can considerably improve the forest structure, quality, and productivity and accelerate forest growth and the generation of more timber, fiber, and other non-timber forest products, which will lead to more materials available for the domestic wood products industry and bioenergy production, as well as greater amounts of carbon to be sequestered and other services provided (Hou et al. 2019). Larger consumption of domestic timber can in turn reduce China's ecological footprints overseas and increase carbon stored in HWPs (Ke et al. 2020, Hou et al. 2019). Together, these shifts will help China to advance toward meeting its national goals of climate change mitigation and ecological civilization.

Under the more aggressive scenario, the added stock volume would total $3.128 \times 10^9 m^3$ for 2021–2030, and the aggregate gain would reach $16.549 \times 10^9 m^3$ over 2006–2060. This stock volume increase would translate into forest biomass and soil carbon additions of 8.463Pg and 4.168Pg. The sum of these amounts of biomass and soil carbon sequestration is 1.322Pg larger than the base case presented in the last section. So, it is critical for China to adopt more effective silvicultural practices. The current annual thinning and actively managed forest area is not only small and slow but also primarily driven by government financing (NFGA 2019).

Greater consumption of domestic wood and fiber resources can increase carbon stored in HWPs, especially if more sawnwood and wood panels can be manufactured and consumed and more fossil fuels and their derivative products and non-renewable construction materials (e.g., cement and steel) can be substituted by wood-based biofuels, bio-energies, or other biomaterials (Gustavsson et al. 2021). Further, post-service HWPs can be recycled and reused in manufacturing reconstituted commodities or making biofuels (Jonsson et al. 2021). These opportunities will go a long way toward meeting the national goals of CC mitigation and ecological civilization.

To improve the efficiency and productivity of its forestry, however, China must further restructure its institutional arrangements and reform its forest policies. In addition to continuing forest tenure and organizational reforms, one priority for the authorities to consider is whether it is wise to ban logging and other silvicultural operations in all the natural forests. Another challenge is to determine whether more than half of the total forestland should be designated for ecological purposes exclusively and thus deprive community collectives or individuals, who own over 40% of the designated forests (Liu et al. 2016), of their commercial management and use rights. Without adequate rights and incentives to manage the forest and dispose of and benefit from cultivated trees, it is hard to imagine that farmers and others will make the needed efforts to enhance forest growth and yields.

The idea of separating commercial uses arbitrarily from environmental uses not only complicates resource management but also hampers the realization of optimal FES services, including carbon sequestration and storage (Liu et al. 2016). Likewise, the nationwide logging quota system is detrimental to operational effectiveness, and the industrial processes of wood products manufacturing should be upgraded and better integrated with resource bases to achieve more efficient utilization of the available resources and greater satisfaction in fulfilling societal needs for multiple ecosystem services (Hou et al. 2019).

*Emissions embodied in commodity trade*

Cross-country carbon flows can be measured from the perspective of consumption or production; different perspectives, however, may lead to starkly varied emission outcomes and thus mitigation responsibilities (Branger and Quirion 2014). Thus, it is worthwhile to investigate issues surrounding GHG emissions embodied in the trade of forest products (Böhringer et al. 2022). The two intertwined issues that have attracted broad attention are carbon leakage (Hong et al. 2022, Hou et al. 2022) and border tax adjustment (Böhringer et al. 2022, Frankel 2022), which I examined in Chapter 6.

China is one of the largest exporters of paper products manufactured mostly from wood fibers, based on intense energy use. Along the deeply connected global value chains (GVCs), the division of labor, and specialization on manufacturing activities have enabled certain countries, mostly advanced economies, to shift the production of some consumer goods, including paper products, to other countries, mainly developing ones. The production-based regime of accounting, however, neglects the impact of the specialization and segmentation of GVCs on $CO_2$ emissions, resulting in large leakages embodied in trade flows, partially thanks to the study of Hou et al. (2022).

## 176  *Summary and outlook*

In addition to the imperative to quantify the carbon flows along the GVCs, Hou et al. (2022) insisted that another driver for China to consider consumption-based carbon accounting would indeed be the fact that the European Union (EU) announced the adoption of a border adjustment mechanism (BAM) by applying the domestic carbon pricing to emissions embodied in imported goods. Using an MRIO model and statistics from the World Input-Output Database, Hou et al. (2022) traced carbon emissions embodied in China's paper trade from 2000 to 2014.

They showed that the amount of $CO_2$ emissions of China's paper industry was estimated to be about 64Mt in 2000, but it rose to 152Mt in 2014. While annual $CO_2$ emissions embodied in its imported finished, as well as intermediate, paper products, it rarely went above and beyond 10Mt during 2000–2014, the amount of annual $CO_2$ emissions embodied in exported finished, as well as intermediate, paper products were around 33Mt in 2000 but increased to 83Mt in 2008 before declining to 47Mt by 2014. As such, China's exports of large quantities of paper products led to high carbon emissions within its borders. However, its emissions were much lower if determined from a consumption perspective. In 2014, for example, the amount of $CO_2$ emissions embodied in the net export was 38.33Mt. In comparison, the US had the largest consumption-based emissions among all the major trading partners of China—12.80Mt.

Naturally, my question was: How would the $CO_2$ emissions embodied in China's whole forest products trade be? With the generous assistance provided by Professor Fangmiao Hou, the lead author of Hou (2022), the answer was: although the gap narrowed over time, the industry's production-based $CO_2$ emissions remained far larger than the consumption-based emissions—114.51Mt in 2014, the last year with all the data available for such an estimation. That amount of leaked emission was over one tenth of China's total $CO_2$ leakage induced by international trade for the same year. A decomposition of the embodied carbon flows of China's forest products trade in 2000 and 2014 also showed that its top three destinations of $CO_2$ emission leakage were the United States, Japan, and Korea.

It can be anticipated that more and more national governments and other domestic and international organizations will look for similar assessments of the causes and effects of inter-regional and international GHG emissions induced by land use and land-use change and flows of traded goods. I am confident that my assessment, along with Hong et al. (2022) and Hou et al. (2022), has opened some important directions for future research.

Nevertheless, consumption-based accounting of land-use emissions or traded goods is not yet required by any national policies or international agreements. The science community ought to accumulate strong evidence and explore the implications of alternative accounting or adjusting regimes (Rodrik 2022, Frankel 2022). Well-crafted policies could help ensure that any additional costs associated with avoiding such land-use change and trade practice would at least initially be borne by (typically more affluent) importing regions. The EU and other advanced economies could prevent negative knock-on effects for developing ones—not only by waiting for lower-income countries to introduce their own carbon taxes (which will be a challenge given their limited administrative capability in the field) but also by supporting those that need the most help to reduce their emissions.

## Other issues of general relevance

### *Carbon finance and funding for forest actions*

Chapter 7 addressed the following two fundamental questions: given the massive global investment anticipated to limit the catastrophic effects of climate change by drastically reducing GHG emissions, how can the necessary mitigation and adaptation actions, including those in the forest sector, be funded, and how have they been financed thus far? A recent study suggests a global increase of $3.5 *trillion* per year from what is spent today (Tyson and Weiss 2022) to transform the energy, transportation, manufacturing, building, and other sectors of the economy, as well as to protect and enhance the capacities of carbon capture and storage through land-use and other practices (IPCC 2022a). Therefore, it is useful to inquire into the essentials of carbon finance and funding for forest sector actions.

I began with a deliberation of the economic rationale of how to reduce GHG emissions and the mechanisms used to achieve that goal through various forms of carbon pricing, and then I overviewed the recent developments in designing and implementing carbon pricing mechanisms, before considering the roles that public and private sectors can play and have played, beyond explicit pricing, in financing ER&R. Next, I delved into forest sector actions by examining the current states and relative costs of different initiatives, alternative means they are funded, and other relevant issues. Finally, I closed this chapter with a discussion of the potential opportunities for future inquiry and implementation.

Overall, the existing evidence indicates that only about 25% of global GHG emissions are subject to any price at all, and 75% of emissions that are subject to a price are less than $10 per ton of $tCO_2e$; and REDD+ and market-based forest project and program finance have accounted for a small fraction (~3%) of the total spent on climate mitigation. Moreover, while promising, the latest developments of carbon pricing remain slow and disappointing, and the financial rewards and penalties continue to be weak in incentivizing polluters to abate their emissions.

It should be emphasized, however, that even though carbon pricing alone may not be adequate to induce change at a pace or on the scale required to meet the PA temperature targets, it is vital to create strong incentives to drive down GHG emissions (Nordhaus 2021, Stern 2006), which can take a variety of forms and must be complemented by well-designed policy (World Bank 2021). The concrete mechanisms include carbon tax, compliance and voluntary emission trading, internal pricing, or financing via issuing bonds or mobilizing public budgetary resources. But care must be taken to ensure that the introduction of carbon pricing is done in a way that not only limits emissions and potential climate disruption but also fosters economic growth and social-ecological sustainability (DeFries et al. 2022).

Fortunately, there are many pricing mechanisms that have been successfully deployed to achieve these objectives, and novel approaches continue to emerge as different mechanisms are strengthened, nested, hybridized, and/or integrated. A combination of them is likely to be more efficient and effective (World Bank 2021). Just like public-private partnership can help drive innovation and investment for a sustainable future (Tyson and Weiss 2022), international cooperation and linkage will reduce distortions in trade and capital flows and aid the efficient ER&R by promoting consistency of actions across countries (Nordhaus 2021).

## *Risk, uncertainty, and social discounting*

Once our attention is focused on the role of carbon sequestration and storage by the forest sector, we enter a drastically different world whereby we face the overarching question of how to mitigate climate change over an extremely long horizon. The notions and tools we used to deal with risk and uncertainty previously could have become obsolete (Stern 2006). So, it is crucial to rethink what risk and uncertainty need to be included in our future work.

More recently, analysts have given attention to the socioeconomic impacts of such forest disturbances as fires, pest outbreaks, diebacks, and hurricanes, which have been intensifying because of climate change. As noted by the IPCC (2020), climate change can exacerbate land degradation processes. Therefore, while it is useful to conduct *ex post* assessments of the socioeconomic impacts of forest disturbances, it is essential to capture the ecological impacts of these disturbances in *ex post* assessments as well. Moreover, it is equally important to carry out integrated economic and ecological assessment of forest disturbances *ex ante* by projecting what damages might happen to the ecological and human systems under given scenarios of climate change. In this context, risk and uncertainty are more about the potential damages caused by climate change (IPCC 2022b).

Furthermore, the combination of these factors—coping with likely severe damages of climate change over a long horizon—forces us to confront the challenge of social discounting in assessing the costs and benefits of climate change mitigation (Rennert et al. 2021). It has been recognized that in this context, a normal market-based discounting regime of commercial investment is not appropriate; indeed, even a modestly high but constant discount rate used in public finance may no longer be suitable (Weitzman 1998, Gollier 2002). Chapter 8 was thus intended to elaborate on how to deal with the risk and uncertainty that the forest sector faces in a world of intense climate change and to elucidate the rationale of and approach to social discounting entailed in public finance for climate actions, including NbSs. Along the way,

I also made my case on whether and how a time value of physical carbon might be considered when analyzing forest sector NbSs.

Regarding government's investments in climate mitigation, the arguments are converging in favor of using such a social discounting scheme—starting with a modest rate and letting it decline gradually as the time horizon becomes longer (Weitzman 2001), which is different from a high and constant rate commonly used in commercial discounting over a relatively short horizon (Gollier and Hammitt 2014). But the existence of different approaches to discounting is not necessarily contradictory; rather, they can be reconciled if we examine different investments in terms of their intrinsic characteristics, including missions, maturities, and domains. Anyway, acting on climate is about transforming our economies, particularly our energy systems. In addition to direct jurisdictional commitments to climate financing, there is a key role for public policy and government action in fostering investment and innovation while complementing roles for the private sector (Songwe et al. 2022).

Finally, my explorations have additional policy implications to the forest sector climate actions. First, it is noteworthy that forest landowners tend to be risk averse in their decision making (Hyde 2012). Coupled with their frequent lack of upfront funds for investing in afforestation, reforestation, or forest management, this calls for businesses, especially large corporations who are committed to ESG (environmental, sustainability, and governance) goals, including net zeros, to provide these landowners with the needed payments at the beginning of their actions (FAO 2016). Doing so will allow smallholders to defray the expenses of stand establishment and treatment, which will likely shore up not just their engagement but also their effective participation in carbon offsetting as partially reflected in the improved forest conditions, in addition to the generation of other co-benefits.

Meanwhile, we can now appreciate public forest ownership and management more thoroughly. According to the latest Global Forest Resource Assessment report (FAO 2020b), as much as 73% of the world's forests are government owned and managed, and a large part of them are used for biodiversity conservation and soil and water protection. Governments are also urged to substantially increase investments in restoring degraded forest and other ecosystems (IPCC 2022a). Thus, public forests have assumed a large role and are expected to assume an even larger one in the provision of various ecosystem services of public good nature.

Of course, some of the services, like soil and water protection, are local or regional public goods, while others, including ER&R and biodiversity conservation, are national or global public goods. Coupled with the potential damage of climate change, this necessitates proper social discounting in assessing the costs and benefits of its mitigation. Furthermore, a more prudent consideration of the question of devolved management of public forests is desirable. While devolving some public forests to communities and other local organizations can improve the incentives and outcomes of management (Yin et al. 2016), caution is warranted in maintaining the continuity, stability, and functionality of public forest management.

## Evaluating the impacts of REDD+ initiatives

Forest resource economists have recently devoted a great amount of attention to evaluating the impacts of REDD+ projects and programs. Among the representative studies are Jayachandran et al. (2017) in Uganda, Roopsind et al. (2019) in Guyana, West et al. (2020) in Brazil, and Groom et al. (2022) in Indonesia. Chapter 9 summarized these empirical analyses and commented on their merits and areas of improvement.

Applications of impact evaluation (IE) methods vary by their underlying assumptions regarding how to identify a counterfactual—what would have happened should the policy intervention not be undertaken—and how to resolve selection bias in estimating the treatment effect. Randomized controlled experiments (RCEs) involve a random assignment of the intervention across the sample; the statuses of treatment and control subjects exhibiting similar preprogram characteristics are then tracked over time (Khandker et al. 2010). RCEs thus have the advantage of avoiding selection bias. In its absence, however, the remaining option is to perform an observational study with such methods as difference in differences and propensity score matching.

Using an RCE, Jayachandran et al. (2017) found that the average tree loss per village between baseline and endline was 9.1% of the tree cover in the control group. The average tree loss in the treatment group was 4.2%—about half of the loss in the control group. Put differently, the program averted 0.326 ha of deforestation per eligible family, with an averted $CO_2$ of 183.5 tons, and the average payment per family was only $37.80 over the two years. The other three IEs are observational studies. Roopsind et al. (2019) showed that the Guyana national REDD+ program reduced tree cover loss by 35% during the implementation period, equivalent to 12.8 million tons (Mt) of avoided $CO_2$ emissions—strong evidence that the program met the additionality criterion of carbon accounting and thus achieving the intended impact of the REDD+ program. The cost of the avoided $CO_2$ emissions was estimated to be $19.53/ton.

But the evidence coming out Indonesia and Brazil is not so encouraging. Groom et al. (2022) reported that Indonesia's moratorium on forestland concessions avoided dryland forest loss no more than 0.03% of all land covered by the moratorium in 2012, rising to 0.22% by 2018—showing no significant evidence that the projects had mitigated forest loss. Likewise, West et al. (2020) concluded that in only four of the projects, deforestation was consistently lower in the project site than in the synthetic control (SC). They also found that the crediting baselines overstated deforestation compared to the estimated counterfactuals. As shown in the projects' ex ante estimates, up to 24.8 million carbon offsets could have been generated by the interventions by 2017 (one unit of offsets is normally 1t$CO_2$). But the Verra Carbon Standards database indicated that only 5.4 million tradable credits from these projects had been certified and made available to offset emissions from various sources by that year.

Notably, all the reviewed cases used geospatial cover data derived from satellite imagery, which is advantageous given their availability and manipulability. The use of high-resolution satellite images by Jayachandran et al. (2017) is innovative, as it can generate timely and accurate, continuous observations that may not come easily from an NFI system given its focus on highly aggregate jurisdictions and low frequency of iteration—typically every 5–10 years. However, it should be made clear that if the spatial resolution is coarse (say, one pixel covers $\geq$ 1km $\times$ 1km), then the data may not have the precision that a quality IE demands. For one thing, the accuracy of the land-use map derived from the Landsat imagery with a pixel size of 30m $\times$ 30m rarely gets higher than 90% (Shi et al. 2017). Groom et al. (2022) and Roopsind et al. (2019) used the Global Forest Change data with a resolution of 1.2km $\times$ 1.2km. It remains to be seen under what circumstances the geospatial data can be accepted by the international agencies and the science community as a substitute for forest cover information in carbon accounting.

Because of using forest cover as the outcome variable, another issue is that the other carbon pools—litter and below ground biomass, as well as soil—could have been omitted or estimated inaccurately. Surely, an argument can be made that these carbon pools are simply not covered by the bilateral contracts. If so, however, a distinction should be made between the bilateral REDD+ efforts and the PA's requirement for a Party to show how it has achieved its climate mitigation targets. An added difficulty is that at the relevant aggregate level—national or subnational jurisdictions—it is hard to find a comparison group, leading to the need for creating some sort of SC. However, there are alternative pathways along this direction, causing the variation of outcomes assessed by alternative sets of predictors. So, I wonder if it is realistic for the UNFCCC to adopt it and, even if it is, how the Conference of Parties can have a set of rules regarding the data coverage, predictor selection, and estimate conversion.

At the same time, we should not forget the fact that funding for REDD+ initiatives, mostly based on bilateral agreements with Norway and Germany, is limited in amount and poorly linked with the UNFCCC or the principles of carbon accounting. Moreover, many non-REDD+ countries that have been implementing forest sector climate actions have not received much IE attention yet. Regardless, it is more meaningful to focus on national or subnational jurisdictional initiatives, as articulated in Chapter 2. Roopsind et al. (2019) started their analysis with a focus on the national REDD+ program and a clear discussion of the principles of additionality, permanence, and no leakage in efforts of reducing tree cover loss and carbon emissions in Guyana. This seems more appropriate.

West et al. (2020) has well exposed the challenges encountered in voluntary carbon projects. Similarly, the study of Coffield et al. (2021) of climate benefits of California forest carbon offset projects disclosed the difficulties of quantifying the additionality and thus the need for improving both governance design and implementation protocol. Relatedly, credits from the voluntary projects are generally issued after a third-party audit every one to five years. While crucial, the third-party involvement in auditing and verification can be problematic and thus must be carefully monitored (Zadek 2023).

In short, with appropriate framing and technique, forest policy analysts can provide robust evidence of policy impact, enabling decision makers to better understand the interlinkages between human and natural systems in determining the policy outcomes. This knowledge can then be used to guide their actions in mitigating climate change and protecting, restoring, and managing forest ecosystems. Hence, it is exciting to pursue more rigorous IE studies that draw up a quality research design, build datasets with broader coverage and better enumeration, and elucidate the complex relationships across temporal and spatial scales (Abadie 2021, Lan and Yin 2017).

## Closing remarks

The overall goal of this book is to develop a coherent knowledge base regarding the numerous processes and their varied principles and practices for the reduction of GHG emissions and the enhanced removal of these emissions in the forest sector. To that end, I provided the policy background for discussing the forest sector's role in climate change mitigation, introduced modeling frameworks and techniques for accounting and assessment, gathered illustrative examples from different parts of the world, and summarized the outstanding difficult issues of policy and measurement, among other efforts.

Here are some of the main messages coming out of this book:

- The Paris Agreement's NDC architecture dictates the governance of climate solutions and the associated matters of carbon accounting, progress tracking, stock taking, and impact evaluation.
- It is crucial to carefully understand and practice jurisdictional approaches to climate governance, which are better positioned to facilitate scaling up forest sector actions despite their inherent limitations.
- Forest carbon economics deals with accounting and assessment at different levels of aggregation, to which there exist alternative modeling systems based directly or indirectly on forest inventory data.
- The amount of carbon sequestered in a forest ecosystem or dispersed trees is reflected in biomass, both above and below ground, and in soil, some of which have not been well inventoried to be accounted for.
- Carbon storage in harvested wood products and its differential durations affect the permanence and ultimately additionality of the capacity of carbon removal by a forest that produces timber and carbon jointly.
- Carbon leakage induced by international trade can be determined from the perspective of production or consumption, but consumption-based determination is more consistent with the UNFCCC thesis of "common but differentiated responsibilities."

*Summary and outlook* 183

- To boost carbon sequestration and storage, countries must further restructure their institutional arrangements and policies to improve the efficiency, productivity, and sustainability of their forestry.
- Strong incentives can take a variety of forms to drive down GHG emissions, but the carbon pricing schemes currently have very limited reach in terms of coverage, level, and effect.
- As far as climate change is concerned, risk and uncertainty are more about the potential damages and thus human society's insufficient commitments to reducing and removing GHG emissions, which call for integrated assessments and adaptive strategies.
- Unlike private discounting of commercial investments, coping with harms of climate change over a long horizon requires the adoption of a social discounting rate in assessing the costs and benefits of its mitigation.
- In estimating the treatment effect, a quality impact evaluation of forest carbon policy should construct a counterfactual and resolve selection bias consistent with the NDC's jurisdictional orientation.

I hope that the readers will find these messages relevant and powerful and, thus, my book has opened a whole new field of academic inquiry—global forest carbon policy, economics, and finance. Certainly, much more can and should be done in advancing the research agenda that I have laid out in this book. For one thing, forest sector climate solutions are complicated and challenging. While we have a general knowledge of the cost competitiveness of forest sector actions (Austin et al. 2020), for instance, their local manifestation remains elusive in many parts of the world. More attention must be devoted to understanding the cost effectiveness of these actions (Griscom et al. 2017, Busch et al. 2019). In any case, it is vital to protect forests, especially those primary natural forests in the tropics, from degradation and destruction; meanwhile, Parties must do more to improve their forest conditions as reflected in the structure, quality, and growth by adopting adequate silvicultural practices of reforestation, afforestation, and forest management.

Also, how to link forest sector actions with other means and measures of climate change mitigation is still less clear, which has to do with enhanced funding for these actions, as well as improved ecosystem conditions. As part of an integrated strategy, forest products manufacturing should be more effectively connected with forest resource management so that the forest sector can live up to its expected role in offsetting and ultimately neutralizing carbon emissions on a more efficient and sustainable basis. Surely, tenure reforms and institutional adjustments are a necessarily part of these efforts (Parrotta et al. 2022). In the end, efficient and sustainable forest sector climate solutions will also generate co-benefits and improve the livelihoods of people in many regions of the world (IPBES 2019).

My analysis also suggests a strong desire for reorienting the inquiry into the issues of governing and evaluating forest sector climate actions from the predominantly disaggregated, project-based paradigm to a more coherent, program-oriented one, such that it will be more congruent with and conducive to delivering

## 184  *Summary and outlook*

the NDC targets (Wang et al. 2021). When integrating REDD+ and other forestry efforts into NDCs, however, actions in the LULUCF arena have unique characteristics that must be carefully considered. Thus, it is crucial for researchers and practitioners in economics, ecology, and other disciplines to broaden their analytic scope and toolkit by embracing both carbon accounting and assessment.

## Note

1 See a definition of REDD+ in the Introduction.

## References

Abadie, A. 2021. Using synthetic controls: feasibility, data requirements, and methodological aspects. *Journal of Economic Literature* 59 (2): 391–425; https://doi.org/10.1257/jel.20191450.

Austin, K.G., Baker, J.S., Sohngen B., Wade, C.M., Daigneault, A., Ohrel, S., Ragnauth, S., Bean, A. 2020. The economic costs of planting, preserving, and managing the world's forests to mitigate climate change. *Nature Communications* 11 (1): 1–9.

Böhringer, C., Fischer, C., Rosendahl, K.E., Rutherford, T.F. 2022. Potential impacts and challenges of border carbon adjustments. *Nature Climate Change* 12: 22–29.

Branger, F., Quirion, P. 2014. Would border carbon adjustments prevent carbon leakage and heavy industry competitiveness losses? Insights from a meta-analysis of recent economic studies. *Ecological Economics* 99: 29–39.

Busch, J., Engelmann, J., Cook-Patton, S.C., Griscom, B.W., Kroeger, T., Possingham, H., Shaymsundar, P. 2019. Potential for low-cost carbon dioxide removal through tropical reforestation. *Nature Climate Change* 9: 463–466.

Coffield, S.R., Vo, C.D., Wang, J.A., Badgley, G. et al. 2022. Using remote sensing to quantify the additional climate benefits of California forest carbon offset projects. *Global Change Biology* 28: 6789–6806. https://doi.org/10.1111/gcb.16380.

DeFries, R., Ahuja, R., Friedman, J., Gordon, D.R., Hamburg, S.P., Kerr, S., Mwangi, J., Nouwen, C., Pandit, N. 2022. Land management can contribute to net zero. *Science* 376 (6598): 1163–1165.

Duan, H., Zhou, S., Jiang, K., Bertram, C., Harmsen, M., Kriegler, E., van Vuuren D.-P., et al. 2021. Assessing China's efforts to pursue the 1.5°C warming limit. *Science* 372: 378–385.

Ecosystem Marketplace. 2021. *State of Forest Carbon Finance 2021.* Washington DC: Forest Trends Association.

Ekholm, T. 2016. Optimal forest rotation age under efficient climate change mitigation. *Forest Policy and Economics* 62: 62–68.

Food and Agriculture Organization of the United Nations (FAO). 2016. *Forestry for a Low-Carbon Future: Integrating Forests and Wood Products in Climate Change Strategies* (FAO Forestry Paper 177). Rome, Italy.

Food and Agriculture Organization of the United Nations (FAO). 2020a. *Trees Outside Forests: Towards a Better Awareness.* Rome, Italy.

Food and Agriculture Organization of the United Nations (FAO). 2020b. *Global Forest Resource Assessment Report.* Rome, Italy.

Frankel, J. 2022. Let the WTO referee carbon border taxes. *Project Syndicate* (November 29).

Fuss, S., Lamb, W.F., Callaghan, M.W., Hilaire, J., et al. 2018. Negative emissions—part 2: Costs, potentials and side effects. *Environmental Research Letters* 13 (6); https://doi.org/10.1088/1748-9326/aabf9f.

Gollier, C. 2002. Discounting an uncertain future. *Journal of Public Economics* 85: 149–166.

Gollier, C., Hammitt, J.K. 2014. The long-run discount rate controversy. *Annual Review of Resource Economics* 6 (1): 273–295; http://dx.doi.org/10.1146/annurev-resource-100913-012516.

Grassi, G., Stehfest, E., Rogelj, J., van Vuuren, D., Cescatti, A. et al. 2021. Critical adjustment of land mitigation pathways for assessing countries' climate progress. *Nature Climate Change* 11: 425–434.

Griscom, B.W., Adams, J., Ellis, P.W., Houghton, R.A., Lomax, G., Miteva, D.A., Schlesinger, W.H., Shoch, D., Siikamäki, J.V., Smith, P., et al. 2017. Natural climate solutions. *PNAS* 114 (44): 11645–11650.

Groom, B., Palmer, C., Sileci, L. 2022. Carbon emissions reductions from Indonesia's moratorium on forest concessions are cost-effective yet contribute little to Paris pledges. *PNAS* 119 (5): e2102613119.

Gustavsson, L., Nguyen, T., Sathre, R., Tettey, U.Y.A. 2021. Climate effects of forestry and substitution of concrete buildings and fossil energy. *Renewable and Substantiable Energy Review* 136 (110435).

Hartman, R. 1976. The harvesting decision when a standing forest has value. *Economic Inquiry* 14 (1): 52–58.

Hoel, M., Holtsmark, B., Holtsmark, K. 2014. Faustmann and the climate. *Journal of Forest Economics* 20 (2): 192–210.

Hong, C.P., Zhao, H.Y., Qin, Y., Burney, J.A., Pongratz, J., Hartung, K., Liu, Y., Moore, F.C., Jackson, R.B., Zhang, Q., Davis, S.J. 2022. Land-use emissions embodied in international trade. *Science* 376: 597–603.

Hou, F.M., Su, H.Y., Liu, C., Lin, X.X., Zuo, F.Y., Xiao, H. 2022. Carbon emissions embedded in China's paper trade: Estimated outcomes of alternative approaches. *Forest Policy and Economics* 145 (102863).

Hou, J.Y., Yin, R.S. 2022. How significant a role can China's forest sector play in decarbonizing its economy? *Climate Policy* 23 (2): 226–237. doi:10.1080/14693062.2022.2098229.

Hou, J.Y., Yin, R.S., Wu, W.G. 2019. Intensifying forest management in China: What does it mean, why, and how? *Forest Policy and Economics* 98: 82–89.

Hyde, W.F. 2012. *The Global Economics of Forestry*, 1st ed. New York: RFF Press/Routledge.

Imbens, G.W., Rubin, D.B. 2015. *Causal Inference for Statistics, Social, and Biomedical Sciences.* New York: Cambridge University Press.

International Panel on Biodiversity and Ecosystem Services (IPBES). 2019. *The assessment report on Land Degradation and Restoration: Summary for Policymakers.*

International Panel on Climate Change (IPCC). 2006. *Guidelines for National Greenhouse Gas Inventories.* Available at: www.ipcc.ch/report/2006-ipcc-guidelines-for-national-greenhouse-gas-inventories/.

International Panel on Climate Change (IPCC). 2020. *Climate Change and Land: Summary for Policymakers.* Available at: www.ipcc.ch/srccl/.

International Panel on Climate Change (IPCC). 2022a. *The Sixth Assessment Report .* Available at: www.ipcc.ch/assessment-report/ar6/.

International Panel on Climate Change (IPCC). 2022b. *Climate Change 2022: Impacts, Adaptation and Vulnerability*. Working Group II Contribution to the Sixth Assessment Report.

Jiang, F., Chen, J.M., Zhou, L.X., Ju, W.M., Zhang, H.F., Machida, T., Ciais, P., Peters, W., Wang, H.M., Chen, B.Z., Liu, L.X., Zhang, C.H., Matsueda, H., Sawa, Y. 2016. A comprehensive estimate of recent carbon sinks in China using both top-down and bottom-up approaches. *Nature Scientific Reports* 6 (22130).

Johnston, C.M.T., Radeloff, V.C. 2019. Global mitigation potential of carbon stored in harvested wood products. *PNAS* 116: 14526–14531.

Jonsson, R., Rinaldi, F., Pilli, R., Fiorese, G., Hurmekoski, E., Cazzaniga, N., Robert, N., & Camia, A. 2021. Boosting the EU forest-based bioeconomy: Market, climate, and employment impacts. *Technological Forecasting & Social Change* 163 (120478).

Jayachandran, S., de Laat, J., Lambin, E.F., Stanton, C.Y., Audy, R., Thomas, N.E. 2017. Cash for carbon: A randomized trial of payments for ecosystem services to reduce deforestation. *Science* 357: 267–273.

Ke, S.F., Qiao, D., Yuan, W.T., He, Y.J. 2020. Broadening the scope of forest transition inquiry: What does China's experience suggest? *Forest Policy and Economics* 118 (102240).

Khandker, S.R., Koolwal, G.B., Samad, H.A. 2010. *Handbook on Impact Evaluation: Quantitative Methods and Practices*. Washington, DC: The World Bank.

Lan, J., Yin, R.S. 2017. Policy impact evaluation: Future contributions from economics. *Forest Policy and Economics* 83: 142–145.

Li, X.Y., Lu, G., Yin, R.S. 2020. Research trends: Adding a profit function to forest economics. *Forest Policy and Economics* 113 (102133).

Liu, P., Yin, R.S., Li, H. 2016. China's forest tenure reform and institutional change at a crossroads. *Forest Policy and Economics* 72: 92–98.

Nabuurs, G.J., Hengeveld, G.M., van der Werf, D.C., Heidema, A.H. 2010. European forest carbon balance assessed with inventory based methods—An introduction to a special section. *Forest Ecology and Management*. doi:10.1016/j.foreco.2009.11.024.

National Forestry and Grassland Administration of China (NFGA). 2019. *A Summary of the China's Forest Condition Derived from the Ninth National Inventory*. Beijing, China.

Nordhaus, W.D. 2019. Climate change: The ultimate challenge for economics. *American Economic Review* 109, 1991–2004. https://doi.org/10.1257/aer.109.6.1991.

Nordhaus, W.D. 2021. *The Spirit of Green: The Collisions and Contagions in a Crowed World*. Princeton, New Jersey: Princeton University Press.

Ohrel, S. 2019. Policy perspective on the role of forest sector modeling. *Journal of Forest Economics* 34: 187–204.

Ostrom, E. 2010. Beyond markets and states: Polycentric governance of complex economic systems. *American Economic Review* 100: 641–672.

Pachauri, S., Pelz, S., Bertram, C., Kreibiehl, S., Rao, N.D., Sokona, Y., Riahi, K. 2022. Fairness considerations in global mitigation investments. *Science* 378: 1057–1059.

Parrotta, J., Mansourian, S., Wildburger, C., Grima, N. (eds.). 2022. *Forests, Climate, Biodiversity and People: Assessing a Decade of REDD+*. IUFRO World Series Volume 40. Vienna.

Piao, S.L., Fang, J.Y., Ciais, P., Peylin, P., Huang, Y., Sitch, S., Wang, T., et al. 2009. The carbon balance of terrestrial ecosystems in China. *Nature* 458: 1009–1013.

Rennert, K., Prest, B.C., Pizer, W.A., Newell, R.G., Anthoff, D., Kingdon, C., Rennels, L., Cooke, R., Raftery, A.F., Ševčíková, H., Errickson, F. 2021. *The Social*

*Cost of Carbon: Advances in Long-Term Probabilistic Projections of Population, GDP, Emissions, and Discount Rates* (Brookings Papers on Economic Activity).

Rodrik, D. 2022. Climate before trade. *Project Syndicate* (December 13).

Roe, S., Streck, C., Obersteiner, M., Frank, S., Griscom, B., Drouet, L., et al. 2019. Contribution of the land sector to a 1.5°C world. *Nature Climate Change* 9: 817–828. doi:10.1038/s41558-019-0591-9.

Roopsind, A., Sohngen, B., Brandt, J. 2019. Evidence that a national REDD+ program reduces tree cover loss and carbon emissions in a high forest cover, low deforestation country. *PNAS* 116 (49): 24492–24499.

Shi, M.Y., Yin, R.S., Lv, H.D. 2017. An empirical analysis of the driving forces of forest cover change in northeast China. *Forest Policy and Economics* 78: 200–209.

Sohngen, B., Mendelsohn, R., 2003. An optimal control model of forest carbon sequestration. *American Journal of Agricultural Economics* 85: 448–457.

Songwe, V., Stern, N., Bhattacharya, A. 2022 *Finance for Climate Action: Scaling Up Investment for Climate and Development*. London: Grantham Research Institute on Climate Change and the Environment, London School of Economics and Political Science.

Stern, N. 2006. *The Economics of Climate Change: The Stern Review*. Cambridge University Press. Available at: http://mudancasclimaticas.cptec.inpe.br/~rmclima/pdfs/destaques/sternreview_report_complete.pdf.

Tyson, L., Weiss, D. 2022. Public-private decarbonization. *Project Syndicate* (April 22).

United Nations Framework Convention on Climate Change (UNFCCC) Secretariat. 2020. *Reference Manual for the Enhanced Transparency Framework under the Paris Agreement*. Bonn, Germany.

Van Kooten, G.C., Binkley, C.S., Delcourt, G. 1995. Effect of carbon taxes and subsidies on optimal forest rotation age and supply of carbon services. *American Journal of Agricultural Economics* 77 (2): 365–374.

Von Essen, M., Lambin, E.F. 2021. Jurisdictional approaches to sustainable resource use. *Frontiers in Ecology and the Environment* 19 (3): 159–167. doi:10.1002/fee.2299.

Wang, Y., Li, L., Yin, R., 2021. A primer on forest carbon policy and economics under the Paris Agreement: Part I. *Forest Policy and Economics* 132 (102595).

Weitzman, M.L. 1998. Why the far-distant future should be discounted at its lowest possible rate. *Journal of Environmental Economics and Management* 36 (3): 201–208. https://doi.org/10.1006/jeem.1998.1052.

Weitzman, M.L. 2001. Gamma discounting. *American Economic Review* 91 (1): 260–271.

West T.A.P., Borner J., Sills E.O., and Kontoleon A. 2020. Overstated carbon emission reductions from voluntary REDD+ projects in the Brazilian Amazon. *Proceedings of the National Academy of Sciences* 117 (39): 24188–24194.

World Bank. 2021. *State and Trends of Carbon Pricing*. Washington, DC.

Wunder, S., Duchelle, A.E., Sassi, Cd., Sills, E.O., Simonet, G., Sunderlin, W.D. 2020. REDD+ in theory and practice: How lessons from local projects can inform jurisdictional approaches. *Frontiers in Forests and Global Change* 3:11. doi:10.3389/ffgc.2020.00011.

Yin, R.S., Pienaar, L.E., Aronow, M.E. 1998. The productivity and profitability of fiber farming. *Journal of Forestry* 96 (11): 13–18.

Yin, R.S., Zulu, L., Qi, J.G., Freudenberger, M., Sommerville, M. 2016. Empirical linkages between devolved tenure systems and forest conditions: challenges, findings, and recommendations. *Forest Policy and Economics* 73: 294–299.

Zadek, S. 2023. Trouble with carbon markets. *Project Syndicate* (March 1).

# Index

1.5°C increase limit 1, 51

Abadie, A. 155
accountability 4, 13, 32–33, 35, 117–118, 127, 160–161
accounting 2–6, 13, 16–19, 25, 27–28, 31–32, 41–56, 78–80, 113, 162–163, 167–168, 170, 181–182, 184; additionality *see* additionality; alternative methodologies 44–46; case studies 50–55; direct approach 5, 41–42, 45–46, 49, 52, 55–56, 170–171; indirect approach 5, 41–42, 45, 47, 49–50, 55, 170–171; in individual countries *see individual countries*; leakage *see* leakage; leakage induced by trade 98–110; local-level frameworks 61–75, 173–174; measurement *see* carbon measurement; permanence *see* permanence; strengths of different approaches 46–50
adaptation 1, 7, 9, 13, 14, 28, 35, 113, 122, 132, 134, 141, 168, 177
additionality 2, 4, 6, 9, 13, 17–20, 26–27, 29, 32–33, 43, 51, 53, 63–65, 70, 72–73, 79, 91, 123, 125–127, 157, 162, 168–169, 180–182
afforestation 3, 16, 18, 27, 30–31, 47, 50, 79–80, 123–124, 144–145
agriculture 1, 5, 32, 34, 90, 94, 97, 101, 117, 122, 126, 167, 175
Amazon (company) 28, 119
Amazon (rainforest) 34, 122, 150, 158–159
American Carbon Registry 119
Andam, K.S. 149
Anti Carbon Project 34
anthropogenic testing 15–16
Argentina 101
Asante, P. 63

assessment xvi, 2, 3, 4, 9, 14, 16, 17, 18, 28, 33, 41–60, 61, 63, 64–65, 80, 81, 88, 90–91, 110, 113, 132, 133–134, 137, 167–168, 170–172, 178, 182
Athey, S. 151
Austin, K.G. 49
Australia 3, 101
Austria 103

Bali Action Plan 3
barriers to entry 31, 34–35, 48
barriers to trade 105
Belgium 103
bioenergy 1, 47, 49, 52, 85, 89–90, 167, 174–175
biofuels 74, 89–90, 149, 173, 175
biomass 2, 4–5, 14–16, 19, 44, 46, 48, 51–52, 54, 64, 68, 80–81, 83, 86–88, 90–91, 95–96, 160, 172, 174, 181–182
Böhringer, C. 99–100, 103–104
border adjustment mechanisms 99–100, 102, 104–105, 110, 176
border carbon adjustments 103–106
border taxes 6, 99–100, 102, 104–105, 110, 176
Boss, D.A. 28
Boyd, W. 9
Brandão, F. 28
Branger, F. 97, 99, 111, 175, 184
Brazil 3, 17, 34, 101, 103, 122, 150, 158–159, 163, 180
Busch, J. 88
business as usual scenario 16, 18, 29, 34, 123, 158, 161–163

California Air Resources Board 119
Canada 5, 14, 50, 52–53, 101, 103, 132, 172
Canadian Forest Service 52–53

Index  189

cap and trade 115, 117–119
carbon accounting *see* accounting
carbon assessment 41–56, 61, 75, 167–168, 170, 182, 184
Carbon Budget Model 52
Carbon Budget Modeling Framework for Harvested Wood Products 52
carbon credits 18, 25, 28–29, 31, 33–34, 36, 72–74, 118–120, 123, 143, 157, 159–160, 163, 180–182
carbon fertilization 45–46, 50–51, 171
carbon leakage *see* leakage
carbon measurement 5, 16, 25, 28, 34, 52, 80–81, 94–96; anthropogenic testing 15–16; reference levels 16–17, 19–20, 29, 79
carbon modelling 5, 44–48, 56, 167, *see also individual models*
carbon neutrality 12, 42, 65, 91, 118, 124–125, 174
carbon offsetting *see* offsetting
carbon peaking 10–11, 82, 94
carbon pricing 6, 20, 30, 99–100, 102–105, 114–127, 133, 176–178, 183; cap and trade 115, 117–119; taxation *see* taxation
Carbon Pricing Leadership Coalition 116, 121
carbon rental 142–143
carbon sequestration and storage 2, 4–6, 10, 14, 18–21, 27, 30, 44, 61, 64–65, 72, 91, 124, 131, 178, 183; in biomass 4–5, 14–16, 19, 44, 80–81, 83, 86–88, 90–91, 95–96, 160, 174, 181–182; costs of 5; in HWPs *see* harvested wood products; in soil 4–5, 14–16, 19, 44–45, 80–81, 83, 86–87, 90–91, 160, 169, 174, 181–182
carbon tax 116–117, 125, 127, 137, 176–177
Center for Climate and Energy Solutions 121
Certified Emission Reductions 63, 65
Chagas, T. 9
Chimpanzee Sanctuary and Wildlife Conservation Trust 156
China 3, 5–6, 14–17, 20, 29, 34, 44, 56, 78–91, 94–96, 103, 118, 122, 124, 141, 154, 173–176; accounting 18–19, 78–80, 98–110
Clean Development Mechanism 3, 27, 30–31
climate bonds 120–121, 127
climate change 1, 7, 131–145, 167, 178, 183

Cline, W.R. 139
Coffield, S.R. 162, 181
Colombia 117
competitiveness 99, 103–105
Conference of Parties 11, 17, 41, 122, 141, 181
corporate social responsibility 119, 121
Costa Rica 117, 149–150
cost-effectiveness 1, 30–31, 61, 88, 99, 103–104, 110, 112, 156, 167, 174, 183
Coulston, J.W. 48–49
COVID-19 pandemic 100
Czech Republic 103

Daigneault, A. 49
dead organic matter 9
deforestation 4, 10, 13–14, 16, 28–29, 34, 51, 53, 122, 149–150, 155, 157–159, 163, 180
DeFries, R. 28–29, 31–32
Democratic Republic of Congo 17, 30
developing countries 3, 6, 13–14, 99–101, 110, 121–122, 141, *see also individual countries*
difference in differences 149, 152–153, 163–164, 180
direct inventory-based approach 5, 41–42, 45–46, 49, 52, 55–56, 170–171
discount rates 7, 135–141, 144, 168, 178–179, 183
drought 132, 134
Duan, H. 80–81, 88, 94, 174
Dynamic Integrated Climate-Economy 137, 139

ecological models 45–46
ecological restoration programs 124
economic growth 47, 85, 95, 97, 117–118, 126, 137–138, 178
Ecosystem Marketplace 31
Ekholm, T. 62
emission reduction and removal (ER&R) 4, 10, 16, 26–29, 31, 35, 41–43, 46, 51, 66, 78–81, 87–88, 95, 114–115, 118–119, 120–121, 123–125, 132, 137–138, 143, 170–171, 174, 177–179
emissions trading schemes *see* cap and trade
emissions-intensive and trade-exposed industries 99, 105
Enhanced Transparency Framework 12
entry barriers 31, 34–35, 48
ESG strategies 119, 121, 127, 144–145, 179

# 190  Index

European Union 5–6, 50, 53–54, 99–100, 102, 104–105, 110, 118, 172, 176–177; Emissions Trading System 100

FAOSTAT database 47, 81, 96
Faustmann model 61–65, 74, 172
Favero, A. 47, 49
financial crisis 107
Finland 103
Fischer, R. 9
flooding 32, 132
FONAFIFO 117
Food and Agriculture Organization 14, 33, 44, 47, 54; FAOSTAT database 47, 81, 96
forest agencies 4–5
forest carbon: accounting *see* accounting; carbon leakage *see* leakage; funding for solutions 3, 6, 26, 28–29, 113–127, 168, 177–178; jurisdictional governance 12–13, 18, 25–37, 65, 168–170; local-level accounting frameworks 61–75, 173–174; policy under Paris Agreement 9–21; program impacts 149–164; risk, uncertainty and social discounting 131–145; role in national decarbonization 78–91, 94–96
forest degradation 2–4, 10, 13–14, 16, 29, 31, 122–123, 145, 155–157, 160–162, 167
forest inventory 14, 33, 42, 45, 48, 50, 52, 55, 67, 132, 159, 170–172
Forest Inventory and Planning 54
Forest Inventory Projection Models 48
forest sector economic models 45, 171
forest tenure 90–91, 149, 163, 175
FORSTAT database 14
fossil fuels 15, 52, 85, 90, 101, 120
framing challenges 160–162, 182
France 103, 136
Frankel, J. 97, 99, 104, 105, 110, 111, 175, 176, 184
free-riding 104
funding 3, 6, 26, 28–29, 113–127, 168, 177–178
Fuss, S. 7, 61, 65, 74, 75, 80, 85, 88, 90, 92, 174, 185

G7 20
Galik, C. 9, 27
Gardeazabal, J. 155
Germany 101, 103, 163, 181
Global Change Analysis Model by Tsinghua University 80, 82–83, 94–95

Global Forest Products Model 47, 96
Global Forest Resource Assessment 145, 179
Global Timber Model 50–51
global value chains 101–102, 175
Gold Standard 119
Gollier, C. 133, 135–136, 140
Governor's Climate and Forests Task Force 25
Grassi, G. 16, 46, 172
green bonds 127
Green Climate Fund 122, 125, 161
greenhouse gas 2–4, 7, 9, 17, 63, 78, 99, 123, 138, 168
Gren, I-M. 9
Griscom, B.W. 1, 8, 9, 20, 22, 24, 38, 39, 41, 57, 59, 76, 92, 122, 129, 132, 146, 148, 167, 183, 184, 185, 187
Groom, B. 150, 158, 160–161, 180–181
Gu, L. 34
Guo, J. 47, 49
Gustavsson, L. 175, 185
Gutrich, J. 63
Guyana 150, 157, 162, 180–181

Hammitt, J.K. 133, 135–136, 140
Hartman modification 61–63, 74, 172
harvested wood products 2, 4–6, 10, 14–16, 19–20, 44–45, 52, 61–67, 72, 74, 78–81, 85–91, 96, 171–175
Henderson, J.D. 132
herbicides 68–71, 73
Hoel, M. 74, 75, 173, 185
Holmes, T.P. 134
Holtsmark, K. 63
Hong, C.P. 100–102, 109–110, 176
Hou, F.M. 100, 102, 106, 110, 175–176
Hou, G.L. 62–63
Hou, J.Y. 49
Houghton, R.A. 50
Howarth, R.B. 63
Hsiang, S. 134–137
hurricanes 132
Hyde, W.F. 62, 65, 76, 144, 146, 179, 185

Imbens, G.W. 151
impact evaluation 7, 149–164, 168, 180–182
improved forest management 16, 29, 123–124, 167
India 3, 14, 17, 29
indigenous communities 13, 117, 125, 150
indirect inventory-based approach 5, 41–42, 45, 47, 49–50, 55, 170–171

indispensable capabilities 33–34
Indonesia 3, 17, 30, 100–101, 103, 110, 150, 157–158, 161, 180
instrumental variable method 153, 164
integrated assessment models 16, 45–46, 137, 171–172
Integrated Policy Assessment model of China 80, 82–83, 86, 94–95
Intergovernmental Panel on Climate Change 11, 14, 18, 43–44, 53, 63, 66, 70, 74, 81, 97, 105, 132–134, 173, 178
international trade 1, 3, 6, 13, 97–110, 167–168, 175–177, 182
Italy 101, 103

Janssens, A. 49, 53
Japan 101, 103, 108–110, 176
Jayachandran, S. 150, 155, 159, 180–181
Jiang, F. 95–96
Johnston, C. 47, 49–50, 81, 96
joint production 2, 62, 64, 66, 167
Jones, J.P.H. 50
jurisdictional finance 123–125
jurisdictional governance 12–13, 18, 25–37, 65, 168–170
Juutinen, A. 9–10

Kaarakka, L. 34
Ke, S.F. 20, 23, 78, 80, 89, 93, 95, 103, 106, 112, 174, 186
Korea 3, 101, 103, 109–110, 176
Kyoto Protocol 2–4, 18, 25, 27, 63, 169; Clean Development Mechanism 3, 27, 30–31

Lambin, E.F. 30
Lan, J. 41, 42, 58, 149, 150, 166, 182, 186
land degradation 28, 32, 132, 178
land rights 36, 170
land use change 13, 44, 47, 110, 125, 176
land use, land use change, and forestry 26–27, 36–37, 51, 169, 184
land-use emissions 100–101, 109, 113
leakage 6, 9, 13, 17–18, 20, 26–27, 79, 81, 91, 96–110, 125–126, 167, 176
Léprière, T.C. 49, 53
Lind, R.C. 139
Liu, P. 14, 20, 21, 23, 24, 44, 89, 90, 91, 93, 94, 95, 175, 186
local-level accounting frameworks 61–75, 173–174
logging 13, 90, 158, 175
Luyssaert, S. 49

marginal elasticity of utility 6
McDermott, C. 27–28, 35
McKinsey & Co. 113
measurement *see* carbon measurement
Mehling, M.A. 31
Mendelsohn, R. 48–51
Mexico 3
Microsoft 28, 119
monitoring, reporting, and verification 12, 25, 28, 31, 34–35, 43, 74, 123, 169, 173
multiregional input-output model 101–102, 176
Murray, B.C. 27

Nabuurs, G.J. 43–44, 49, 53–54
National Academies of Sciences, Engineering, and Medicine 137
national forest inventories 14–16, 33, 44–46, 54, 80, 159–160, 181
National Forestry and Grassland Administration (NFGA) 15, 23, 78, 89, 90, 93, 95, 124, 129, 174, 186
nationally determined contributions 1, 3–4, 9–10, 19–21, 25–26, 31, 35–37, 41–43, 64, 66, 78–79, 87–88, 91, 113–114, 117, 122, 124, 127, 158, 160, 162, 168–169, 182–184; of individual countries *see individual countries*; jurisdictional governance 12–13, 18, 25–37, 65, 168–170; progress tracking 11–12, 43, 182; stock taking 11–12, 43, 182; targets and indicators *see* targets
nature-based solutions 1, 27–29, 31, 33, 37, 41, 61–62, 65, 72, 74, 116, 118, 120, 122–124, 126, 132–133, 144, 173, 178–179
nesting 33, 35, 124
net zero targets 10, 12, 42, 78, 82–83, 89, 105, 113, 119–120, 122, 140–141, 145, 173, 179
Netherlands 103
New Zealand 3, 118
Ning, Z. 134
nitrogen deposition 15, 45–46, 171
Nolte, C. 150
non-governmental organizations 13, 32, 47
non-timber forest products 64, 66, 89, 174
Nordhaus, W.D. 7, 27, 114, 133, 137, 139
Norway 157, 161, 163, 181

offsetting 3, 6, 14, 16, 18–21, 25, 29, 31–32, 66, 70, 72–74, 78–81, 86–89,

91, 120, 124, 142–143, 145, 160, 162, 169, 173–174, 179, 183
Ohrel, S. 42–46, 48, 50, 56, 170–171
Oldfield, E.E. 34
Ostrom, E. 35, 140
*Our World in Data* 98

Pachauri, S. 36, 39, 170, 186
palm oil 158
panel data models 153
Pannell, D.J. 9
paper/paperboard 6, 64, 66, 68, 70, 72–73, 80, 100–103, 173, 176
Paris Agreement 1, 9–21, 41–43, 51, 53, 61–64, 66, 72, 74–75, 79, 88, 97, 104, 113, 115–116, 119, 122, 125, 149, 161, 163, 168, 172, 177, 181–182; background on 2–3; carbon accounting 17–19, *see also* accounting; elaboration on 168–170; forest-level stipulations and requirements 13–16; general stipulations and requirements 10–13; NDCs *see* nationally determined contributions; reference levels 16–17, 19–20, 29, 79; understanding 10–21
Parisa, Z. 141–143
Parrotta, J. 27, 113
Party 4, 10, 11–12, 14–15, 18–19, 43–44, 160, 162, 168, 181
Patriquin, M.N. 132
payment for ecosystem services 155–156
per capita growth rate of consumption 6–7
permanence 2, 9, 13, 18–20, 26–27, 30–32, 34, 43, 51, 53, 63–66, 70, 72–74, 79–80, 125–126, 141–143, 157, 162–163, 169, 173, 181–182
pest outbreaks 132, 178
Pfaff, A. 150
physical carbon, time value of 141–144, 179
Piao, S.L. 44
pine 5, 62, 66–74, 172
Pizer, W.A. 134
Plan to Prevent and Control Amazon Deforestation 34
Poland 103
Portugal 103
progress tracking 11–12, 43, 182
project design documents 123
propensity score matching 149, 153, 163, 180
protected areas 149–150
pulpwood 68–73

QuickBird 156

Radeloff, V.C. 49, 81, 96
Ramsey, F.P. 138, 140
Ramsey approach 6, 138, 140
randomized controlled experiments 150–151, 155–157, 160, 163, 180
REDD+ 1–4, 6, 9, 14, 16–18, 20, 25–33, 36–37, 42, 113–114, 119, 122–126, 149–150, 155, 157–164, 168–170, 177, 180–182, 184
reference levels 16–17, 19–20, 29, 79
reforestation 3, 16, 18, 27, 30–31, 47, 50, 79–80, 123–124, 144–145
Regional Greenhouse Gas Initiative 123
regression discontinuity methods 153–154
renewable energy 110, 120–121
Rennert, K. 133, 136–139
results-based payment xv–xvi, 14, 17–18, 20, 21, 26, 29–30, 33, 91, 125, 169
risk 6, 28, 32, 131–136, 138–139, 144, 168, 178–179, 183
Rodrik, D. 100
Roe, S. 1, 8, 9, 20, 24, 27, 39, 41, 42, 45, 59, 61, 76, 132, 148, 167, 171, 187
Rogoff, K. 121
Roopsind, A. 150, 157, 160, 162, 180
rotation age 54, 61–63, 67–68, 70–72, 74, 131, 172–173
Russia 5, 50, 54–55, 103, 109, 122, 172

Saudi Arabia 101
sawnwood 64, 66, 68, 70, 72–74, 81, 89, 132, 173–174
Schepaschenko, D. 49, 54–55
Sedjo, R. 9
selection bias 152–153, 163–164, 180, 183
Seymour, F. 29
shared socioeconomic pathways 45, 47, 81, 85, 96, 171
silvicultural practices 20, 67, 89–91, 95, 174–175, 183
Sjølie, H.K. 56
Sloping Land Conversion Program 154
Smith, C.E. 43, 49, 52
social cost of carbon 67, 115, 135–139, 141, 143
social discounting 6, 136–141, 144, 168, 178–179, 183
Sohngen, B.L. 9, 48–51
soil carbon 4–5, 14–16, 19, 44–45, 80–81, 83, 86–87, 90–91, 160, 169, 174, 181–182
Songwe, V. 134, 141

South Africa 3
southern pine 5, 62, 66–74, 172
Spain 103
static horizon concept 142
Stern, N. 6–7, 122, 131, 133, 136, 138–139
stock taking 11–12, 43, 182
subsidiarity principle 27, 35
Suichang Carbon Project 34
Sun, H.Y. 72
supply chains 13, 33, 109, 121
Sweden 47, 103, 117
Switzerland 103, 117
synthetic controls 155, 157, 159–161, 163–164, 180–181

targets 10–11, 16, 35, 41, 43, 80, 88, 104, 113, 177, 181; 1.5°C increase limit 1, 51; carbon neutrality 12, 42, 65, 91, 118, 124–125, 174; net zero targets 10, 12, 42, 78, 82–83, 89, 105, 113, 119–120, 122, 140–141, 145, 173, 179
taxation 6, 97, 99, 115, 120, 137, 175; border taxes 6, 99–100, 102, 104–105, 110, 176; carbon tax 116–117, 125, 127, 137, 176–177
technological transfer 104
Thamo, T. 9
timber 2, 16, 21, 45, 47, 50–51, 55, 61–73, 78, 80, 88–89, 131–132, 158, 167, 170–174; harvesting 16, 52, 54, 63, 65–66, 68, 72–73, see also harvested wood products
time preference 6
trade, international 1, 3, 6, 13, 97–110, 167–168, 175–177, 182
tragedy of the commons 104
transaction costs 13, 32, 34–35, 125–126, 143, 160
trees outside of forests 14–15, see also urban trees/forests
Turkey 100
Tyson, L. 113, 121, 130, 177, 178, 187

Uganda 150, 155–156, 180
uncertainty 6, 131–133, 144, 168, 178, 183

United Kingdom 101, 103, 136
United Nations Climate Secretariat 41
United Nations Framework Convention on Climate Change 1–3, 11–12, 17–18, 25–26, 42, 54, 99, 105, 122, 127, 160–161, 163, 169, 181–182
United States 5, 14, 17, 29, 34, 48, 56, 62, 66–74, 98, 101, 103, 108–110, 116, 118–120, 122, 132, 134–135, 137, 162, 172, 176, 181
University of Georgia Plantation Management Research Cooperative 67–68
urban trees/forests 14–15, 169

value chains 119–120; global 101–102, 175
Van der Gaast, W. 9–10
Van Kooten, G.C. 42, 63
Verra Carbon Standards 119, 159, 180
Vietnam 110
voluntary markets 25–29, 32, 36, 74, 119–120, 123, 125–127, 161–162, 169, 173, 178
Von Essen, M. 30

Wang, Y. 43, 65–66, 113
Wear, D.N. 48–49
Weitzman, M.L. 133, 136, 139
West, T.A.P. 150, 159, 161–163, 180–181
wood-based panels 64, 66, 74, 81, 89, 102, 173
World Bank 28, 115, 117–120, 125–126, 177–178
World Induced Technical Change Hybrid model 80, 83, 94–95
World Input-Output Database 102, 176
World Trade Organization 97, 105–107
Wunder, S. 26, 28

Yin, R.S. 49, 67, 88

Zadek, S. 36, 40, 126, 130, 162, 166, 182, 187
Zeng, W.S. 14, 24, 44, 60, 91, 94
Zhai, J. 134